The Face of Venus

THE MAGELLAN RADAR-MAPPING MISSION

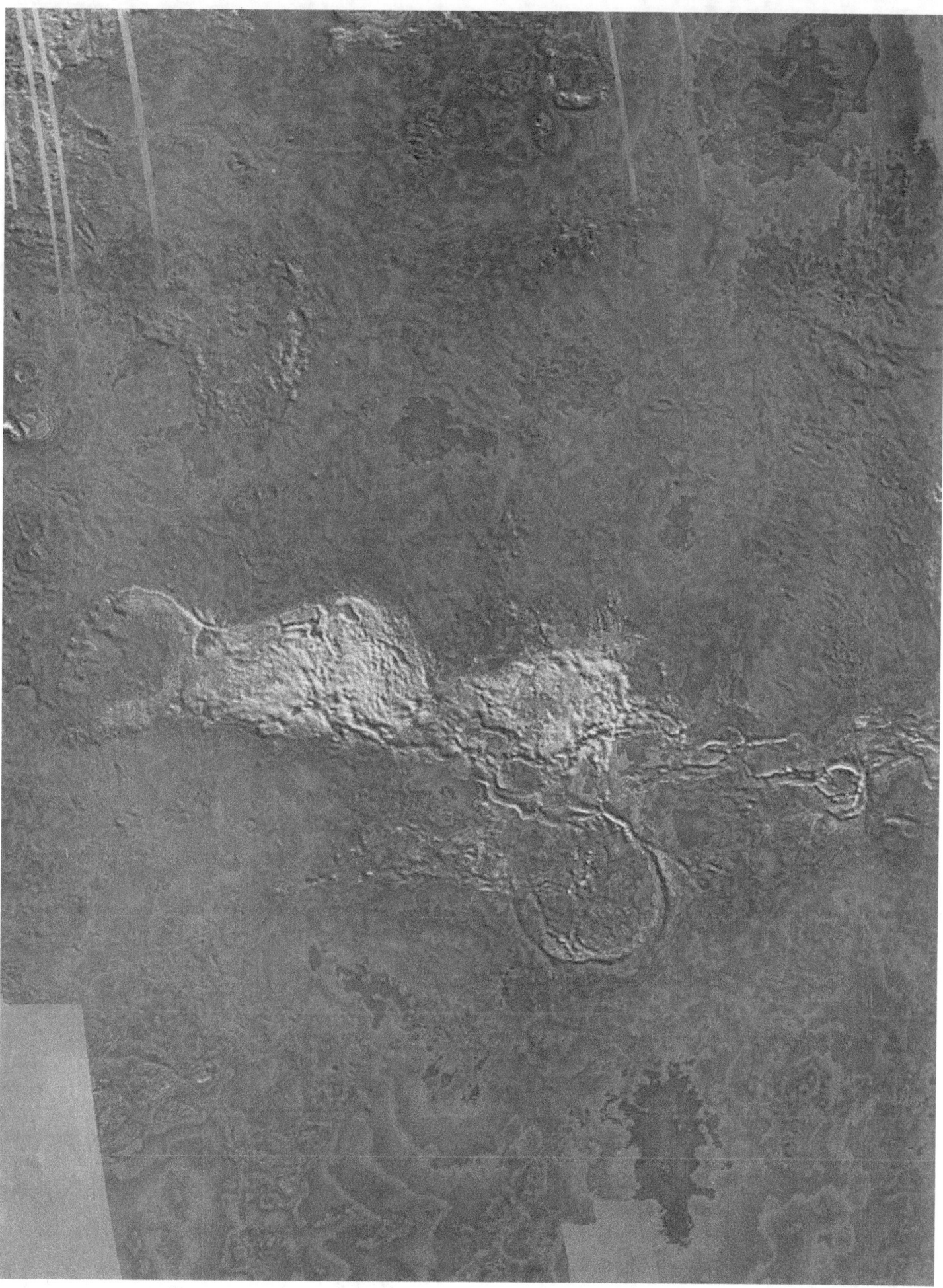

The Face of Venus

THE MAGELLAN RADAR · MAPPING MISSION

Ladislav E. Roth
Stephen D. Wall

Editors

Jet Propulsion Laboratory
California Institute of Technology

NASA SP-520

National Aeronautics and Space Administration
Washington, D.C.

June 1995

Foreword

Magellan, both the sailor and the spaceflight mission, represent in most minds daring adventure, scientific exploration, and vivid story-telling. The adventures of Magellan the sailor — his grit, bravery, and tragic death — could only be captured in words. The story of Magellan the spaceflight — its bold design, resourceful execution, and dazzling success — can fortunately be told in both words and pictures. Telling this story is the purpose of this book.

In the succession of distinguished visitors from Earth to Venus, the Magellan spacecraft has been the most recent. The visitors through the years have consisted of an armada of spacecraft — flybys, orbiters, and landers. Each spacecraft, along with the solitary radar-gazers stationed on Earth, managed to pierce a small hole in the veil of mystery that has surrounded Venus since the discovery of the telescope. Magellan, having imaged almost all of Venus' face, rent the planet's veil to pieces. Many questions about Venus still remain — for instance, why has a planet so similar to Earth followed an evolutionary path so different? And why do conditions on the surface of Venus more closely resemble those in Hades than those on Olympus? Nevertheless, Magellan has provided definitive answers to some of the queries that have motivated the planetary exploration of the last thirty years. These queries address the basic character of venusian geology and offer an assessment of Venus' place within the solar-system family.

The Face of Venus offers but a peek at the treasures to be found in the Magellan dataset — and a glimpse into the volumes that have been and will continue to be written regarding the geology of Venus. As you look through this book, you will be led one-by-one through the treasures Magellan has gathered — in a manner neither stilted nor oversimplified — and you will gain an appreciation of Venus that spans several disciplines and several realms.

Through its Special Publications (SP) program, the National Aeronautics and Space Administration (NASA) has produced a series of books that describe the planetary missions the agency has flown to date. The SP books contain the most compelling images yielded by these missions, accompanied by scientific narrative as well as references to the relevant literature. With the sense of pride inherent in a job well done, the Magellan engineers, operators, and investigators invite you to peruse The Face of Venus — the NASA Special Publication on the results of the Magellan mission.

Wesley T. Huntress, Jr.
Associate Administrator,
Office of Space Science
NASA Headquarters

Contents

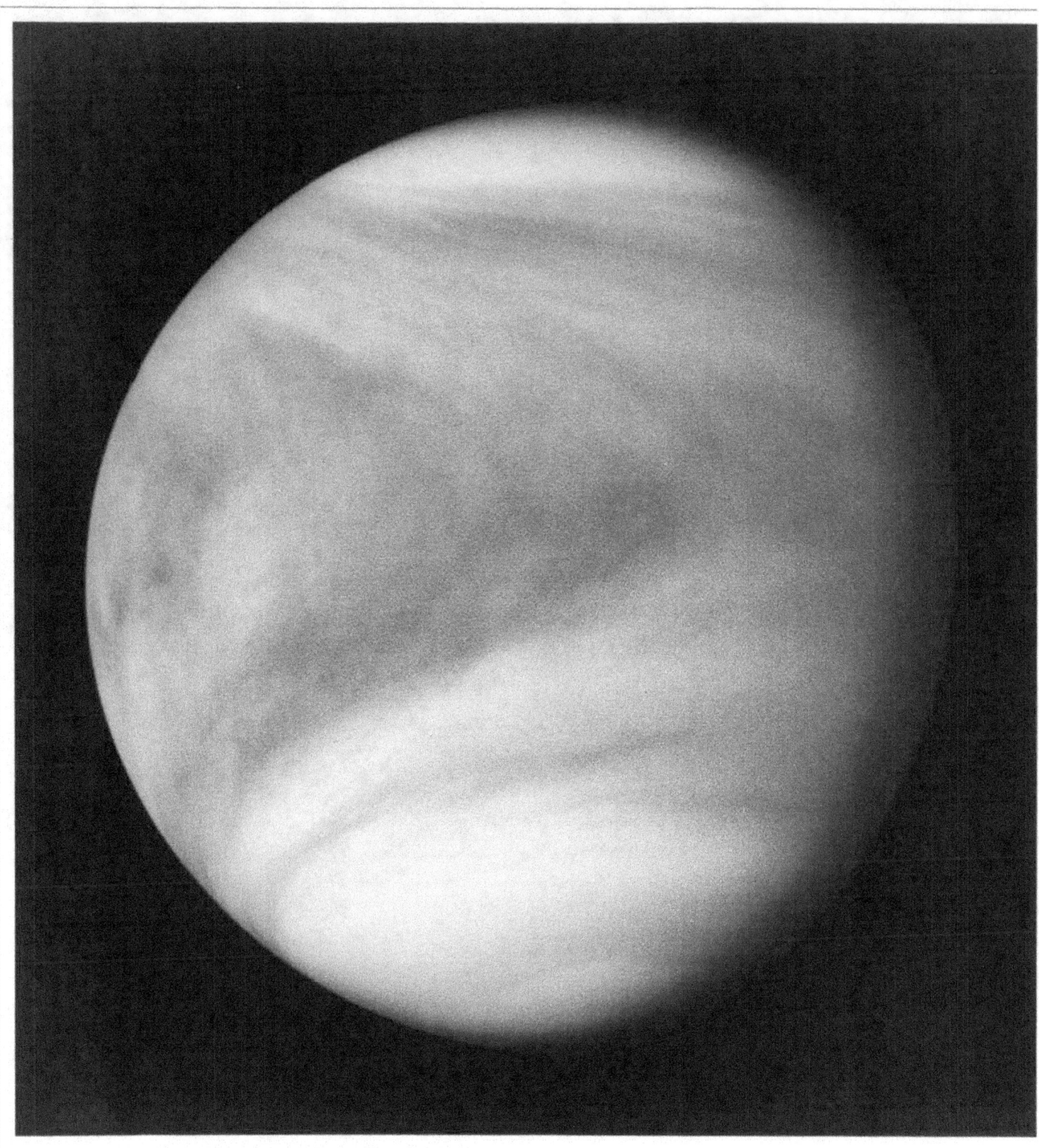

Before Magellan

Venus, brilliant in the early evening or morning sky, one of the seven wandering stars of the ancients, has long been a difficult object to study. Although it is our closest planetary neighbor (only some 40 million kilometers distant at inferior conjunction), until very recently far less has been known about Venus than about the smaller and more distant Mars. Mars, because of the almost total absence of optical obstructions in its thin atmosphere, has been accessible to telescopic observations. Venus, meanwhile, remained secret, mysterious, and hidden behind a dense shroud of clouds.

Image of the disk of Venus, acquired by the Cloud Photopolarimeter instrument aboard Pioneer Venus. The speed of motion of the characteristic Y-shaped cloud patterns across images taken at 24-hr intervals indicates that the topmost atmospheric layers rotate around the planet approximately once every four days. This suggests wind speeds up to 500 km/hr. The moving cloud patterns were also observed in Mariner 10 television images taken through an ultraviolet filter. [R. O. Fimmel, NASA Ames Research Center.] 1.1

The Magellan radar-mapping mission has been but a culmination of mankind's fascination with Venus, which, through the planet's association with the celestial personification of a menagerie of deities, can be traced back to the dawn of recorded history. In the Western view, derived through Rome and Greece from ancient Mesopotamia, the goddess Venus is seen as the embodiment of the feminine principle. The conventions governing the naming of features on the planet's surface, adopted by the International Astronomical Union, reflect that view. In a number of non-Western cultures, the planet has traditionally been associated with male idols. For instance, in pre-Columbian Central America, the various incarnations of the Venus deity Kukulcán-Quetzalcóatl were strictly male. In Egypt, the planet Venus had been associated with Osiris, king of the underworld and symbol of the deceased kings. In China, the planet was known under several apellations — the Grand White, the Metal Planet, the Executioner's Star — none expressly feminine.

The oldest preserved record of the systematic preoccupation with Venus is contained in a Babylonian omen series compiled around 1600 B.C., about a century following the reign of Hammurabi, the fabled king–lawgiver. In succeeding Mesopotamian empires, the risings and settings of Venus constituted a crucial component of the inventory of omens used to foretell the fortunes of the ruler and the state. The accumulated observations of Venus were incorporated in the body of knowledge that formed the foundation of Chaldean and Hellenistic horoscopic astrology — a contribution to civilization that is still with us today.

In the hands of Attic philosophers, the Babylonian veneration of planets turned into a speculative enterprise. Although initially intertwined with astrology, its ultimate goal, after two millenia of effort by astronomers of many nations, came to be the physical understanding of the celestial luminaries — rather than the dim-sighted fear of them. The momentous event during this long, tortuous journey

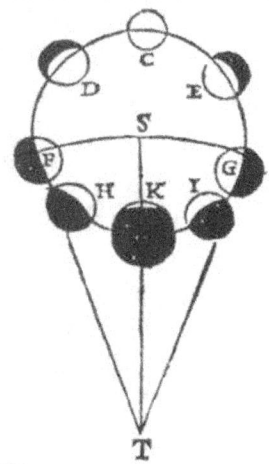

was the announcement in 1543 of the heliocentric hypothesis. In the geocentric world, there was much disagreement regarding the order of the inner planets. Some astronomers thought Mercury and Venus closer to Earth than the Sun; others thought the Sun closer than any other planet. By making the Sun the center of the cosmos, Copernicus solved the ancient problem of the heavenly order and showed Venus to be our closest planetary neighbor. The observational evidence for the new order arrived with Galileo's deft use of the recently invented "spyglass," to show in 1610 that Venus simulated the phases of the Moon.

Further progress in the telescopic studies of Venus was thwarted by the uniformly bright appearance of the planet's crescent. In 1667, after having timed the instants of successive meridional passages of apparent permanent markings on planetary disks, G.-D. Cassini determined the rotation periods of Jupiter, Mars, and Venus. Cassini's periods for Jupiter and Mars were very nearly correct; his Venus period, 23 hours 21 minutes, turned out to be grossly in error. The correct value was not established until after the advent of the radar era.

Sketch of Venus' phases by J. B. Riccioli (*Almagestum Novum*, 1651). Reflecting his period's resistance to the heliocentric conjecture, Riccioli proposed a variant of the geocentric hypothesis — in which Venus, Mercury, and Mars revolved around the Sun, and the Sun, with its retinue, revolved around Earth. This arrangement accounted for the crescentic appearances of Venus without relinquishing Earth's central position in the cosmos. [R. Brashear, Huntington Library.] **1.2**

Chart of alleged permanent markings on the surface of Venus (P. Lowell, 1896). Keeping with the notion of Venus as the goddess of love, Lowell named these altogether fanciful markings after classical mythological figures — female and male — known for their amatory exploits. To the busiest of the intersections of these imaginary markings Lowell assigned the name Eros. **1.3**

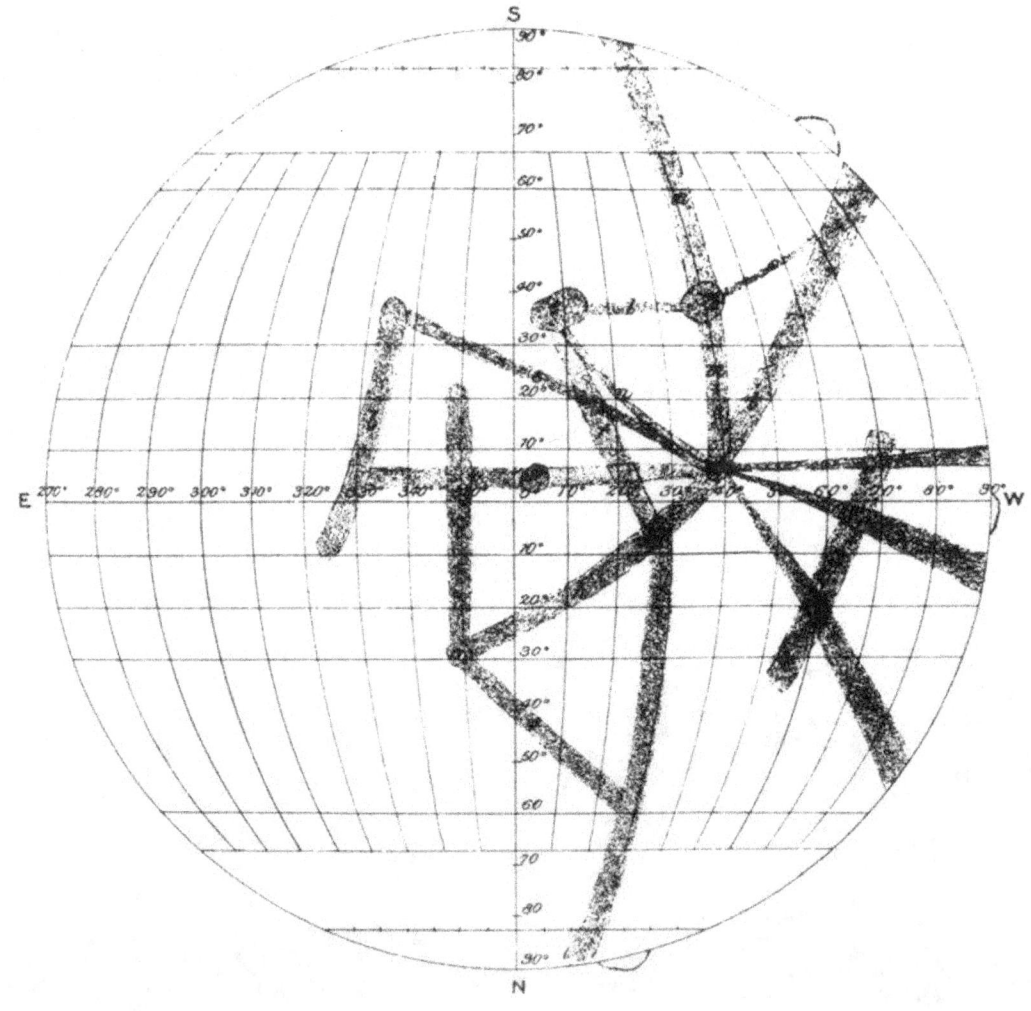

Newly determined rotation periods implied that the lengths of days on the other planets differed little from the length of the terrestrial day. Astronomers concluded that these planets, in this crucial aspect being so similar to Earth, by necessity had to support life not different from that flourishing on Earth. Even C. Huygens, the discoverer of the rings of Saturn and a sharp-eyed observer who rejected Cassini's venusian period by arguing that the Venus crescent was "too brisk" to bear any visible markings, wholeheartedly supported the idea of terrestrial-like life on Venus.

The notion of life on Venus acquired a life of its own and survived, in some quarters at least, well into this century. The gleaming but bland telescopic appearance of Venus led Huygens to suggest that a thick atmosphere might be enveloping the planet. Venus transit observations, carried out by M. V. Lomonosov in 1761 and D. Rittenhouse in 1769, confirmed this suggestion. A realization gradually developed that Venus' extraordinary brightness is not due to the presence of atmosphere alone but to a permanent band of clouds or haze that must be standing between observers on Earth and the planet's surface. Nonetheless, reports on the new determinations of Venus' rotation period — based on sightings of supposedly permanent markings — continued to be published, and some astronomers even claimed that they were able to see through the clouds often enough to produce maps of the venusian surface.

The earliest spectroscopic analyses (1875–1895) of reflected venusian light indicated the presence of free oxygen in the uppermost atmospheric layer. More sophisticated Doppler spectroscopy (1922–1952) wrecked this fancy; instead, carbon dioxide was identified as the dominant atmospheric constituent. However, cloud composition continued to present a vexing problem. As late as 1920, failure to detect water vapor in the atmosphere resulted in presenting Venus as a dry, dusty world. In contrast, by 1955 it was proposed that the clouds consisted almost

Venera 13 panorama of the landscape at the landing site (−7.5°, 303.0°). The angular resolution is about 0.2°. The white object in the middle of the image is the ejected view-port cover. (For scale, the teeth on the circular support are 5 cm apart.) Images taken at Venera landing sites indicate the presence of a loose, fine material, unmarked by the usual telltale signs of eolian action. The imaged material may be of local origin — rather than having been moved to its present position by wind. Also, the material is darker than the rock from which it was derived. (Usually, the pulverized rock is lighter than the parent rock.) [Soviet Academy of Sciences.] 1.4

Note: Coordinates of points on the surface of Venus are given as (lat, lon). "Lat" is the latitude, positive north, negative south; "lon" is the longitude, east only. At the equator, 1° longitude equals 106 km. Scales given in captions are approximate.

exclusively of water vapor. This theory had a curious corollary in that the planet's surface could be covered by steamy marshlands and oceans. One interpretation of Venus radio-emission observations between 1958 and 1970 admitted surface temperatures in the neighborhood of 700 kelvins (~450 degrees Celsius). Polarization studies (1968–1973) showed the clouds to be composed mostly of sulphuric acid. Any visions of a damp, muggy Venus, teeming with aquatic life, now evaporated.

After the Moon, Venus became the next major solar system target to be explored by radar instruments based on Earth. The first echoes bounced off the surface of Venus were received in 1961 by the Goldstone station of NASA's Deep Space Network antenna system. At subsequent inferior conjunctions, Venus was observed by both Goldstone and the National Astronomy and Ionosphere Center's Arecibo Observatory radars. After an optimistic start, attempts by other radio observatories to conduct Venus observations faded away. The ability of the radar signal to penetrate through the clouds to the planet's solid surface made it possible to carry out experiments that were radar analogs to the timing of meridian transits. By 1963, it became evident that the planet's rotation was retrograde, i.e., in the direction opposite to the direction of orbital motion. By 1967, it was established that Venus rotates once in 243.1 days about an axis very nearly perpendicular to the orbital plane. Radar also permitted determination of other physical parameters of relevance to Venus studies. Radius of the planet was set at approximately 6052 kilometers, some 70 kilometers shorter than the radius computed from the telescopically observed apparent diameter, and about 320 kilometers shorter than the mean radius of Earth; the astronomical unit was determined to an accuracy of several kilometers. Furthermore, radar ranging at different positions of the planet's path around the Sun resulted in considerable refinement of the orbital elements.

A dramatic advance in the exploration of Venus, a shift from the planet being seen as a disembodied astronomical object and becoming instead a subject of geologic interest, came with the introduction of radar imaging techniques (1970–1985). Results of the early radar reconnaissance of Venus indicated that the planet's surface rock was more compact than the pulverized surface layer on the Moon. The first

Both Venera 15 and 16 orbiters carried 8-cm-wavelength imaging radar. The chosen orbit geometry permitted imaging at latitudes northward of about 30° latitude. Shown here is the radar mosaic, with topographic contours, of an elevated plateau named Ishtar Terra, in size approximately equivalent to the continent of Australia. The imaged terrain is bounded by latitudes 55° and 80° and longitudes 300° and 360°. Contour separation is 500 m. *[Soviet Academy of Sciences.]* 1.5

Note: In all radar images shown, north is at the top.

Earth-based radar images of Venus, as crude as they were, further highlighted the differences in the physical constitution of the two bodies. These first images permitted identification of a handful of anomalously reflective spots on Venus — features that had no analog on the Moon. (The names Alpha, Beta, and Maxwell were given to those spots and were carried through into the Venus nomenclature system currently in use.) Eventually, with the impressive improvements in the capabilities of Earth-based radars, images of Venus were acquired with resolution as high as 1–2 kilometers.

From the beginning of the space age, Venus was considered a most tempting destination for spacecraft exploration. Opportunities to launch a spacecraft to Venus are spaced at intervals of about 19 months. From 1962 to 1985, almost all those opportunities were utilized. Before landing was attempted, a number of spacecraft were sent to Venus on flyby missions.

The 1962 Mariner 2 flyby of Venus marked the first visit by a human-made object to another planet. In 1965, Venera 3 crashed on Venus, becoming the first craft to reach a planetary surface. In 1967, Venera 4 became the first probe to return data from within the venusian atmosphere, and in 1970, Venera 7 accomplished the first successful landing on Venus. In 1975, Venera 9 transmitted back to Earth the first images from the surface of Venus and conducted the first gamma-ray analysis of landing-site rocks. Mariner 10, the first multiplanet spacecraft, flew past Venus in 1975 on its way to Mercury. In 1978, Pioneer 12 (referred to as Pioneer Venus or Pioneer Venus Orbiter), traveling about Venus in an orbit of considerable eccentricity, compiled the first altimetry and gravity maps of the planet's surface within the strip of lati-

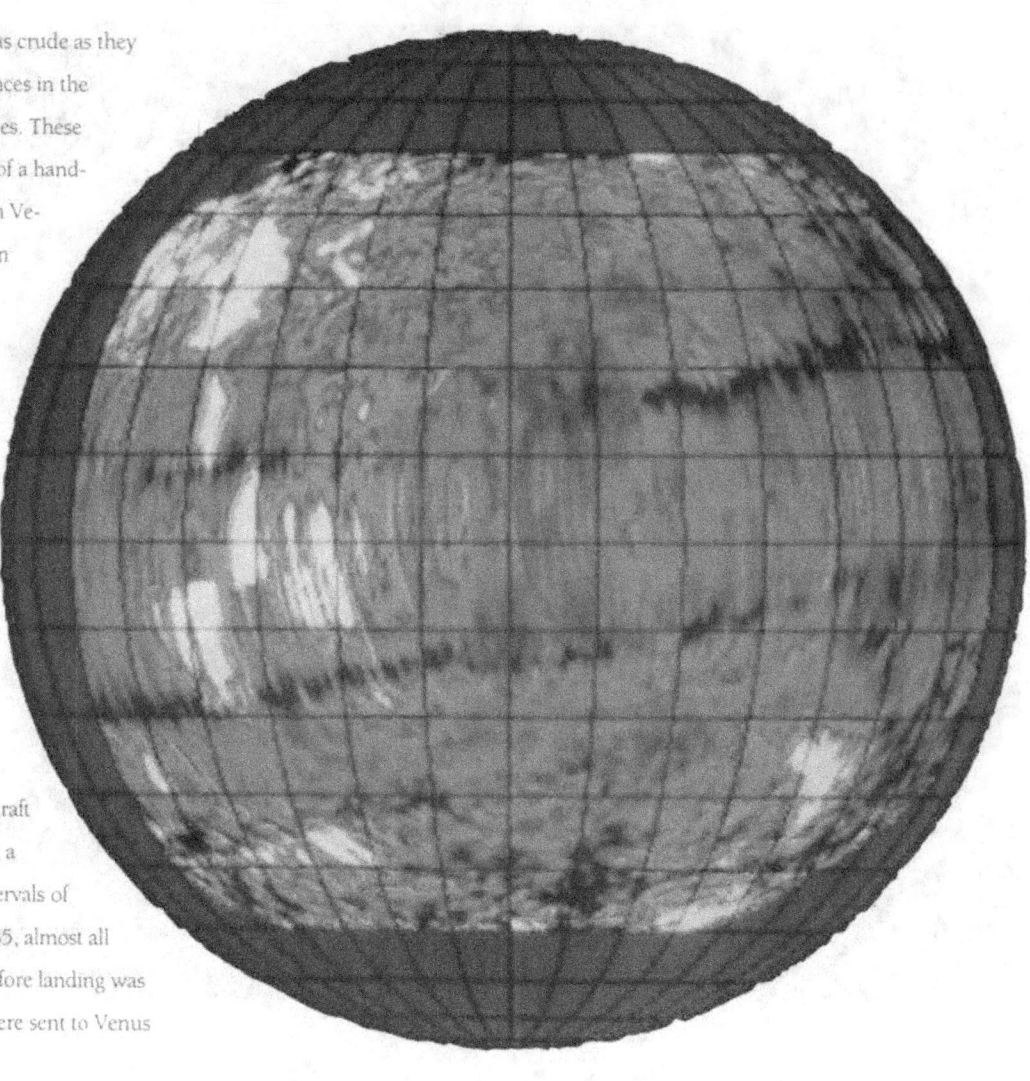

An early (1972) Arecibo 70-cm-wavelength radar interferometer image, with average resolution of about 100 km, of the disk of Venus. Radar-bright Alpha (lower right) and Beta (upper left) regions are plainly visible. [D. B. Campbell, Cornell University and Arecibo Observatory]
1.6

From the beginning of the spaceage, Venus has been a most tempting destination for spacecraft exploration

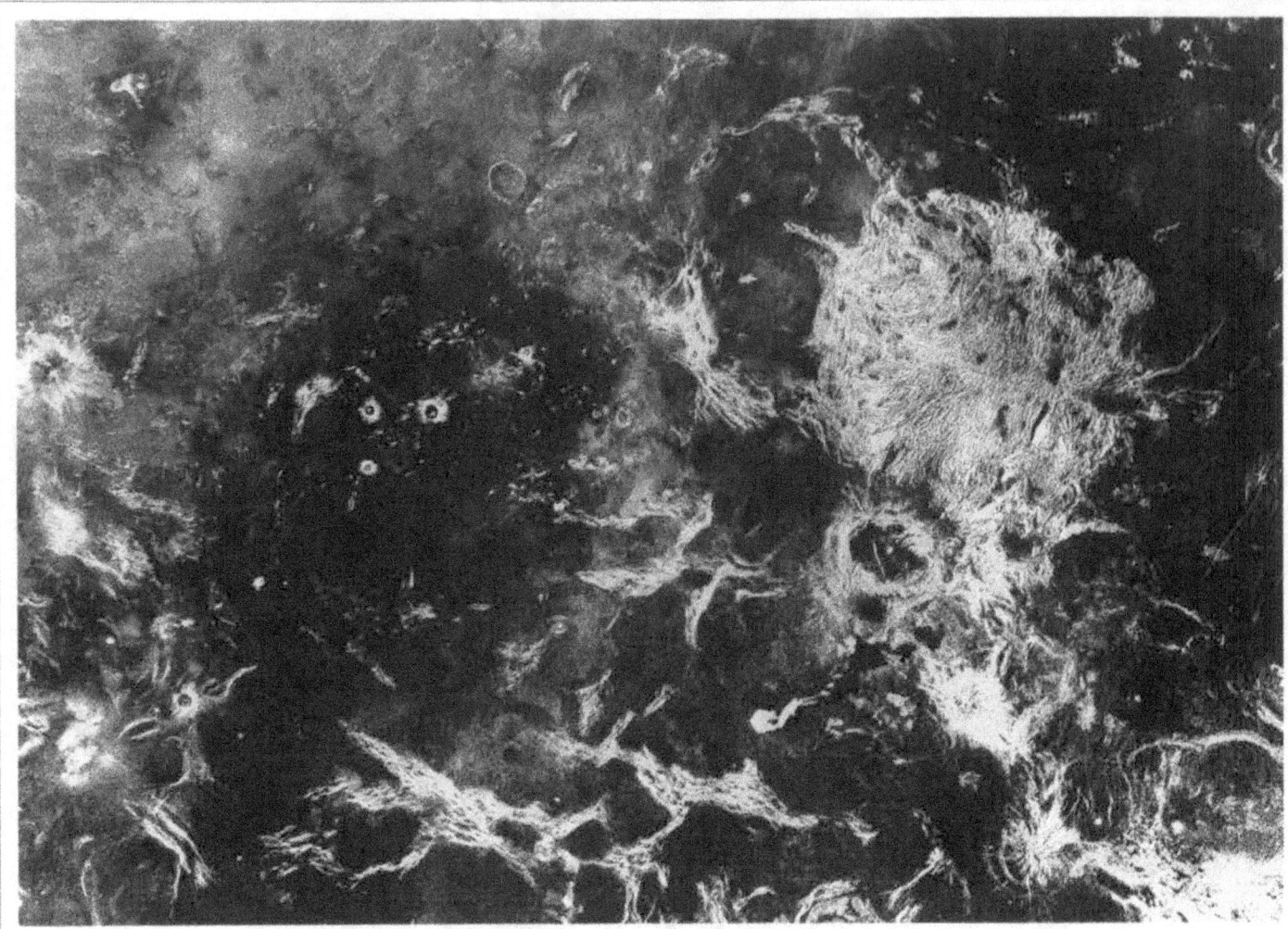

tudes from 78 to –63 degrees. The footprint size of the altimetry data was about 100 kilometers, with vertical accuracy about 100 meters.

Pioneer 13 released four probes into the venusian atmosphere in 1978. Combined, the temperature and pressure surveys by the Venera 8–12 landers and the Pioneer large probe yielded 735 ± 3 kelvins (~460 degrees Celsius) and 90 ± 2 bar (~90 atmospheres) as the measurements at the planet's surface. The in situ temperature measurements thus confirmed the earlier estimates obtained from the radio-emission observations. In 1981, Venera 13 returned the first color images from a landing site on Venus and carried out an X-ray fluorescence analysis on a drilled sample. The lander instruments operated for a record 127 minutes in Venus' searing heat. Also in 1981, the Venera 14 lander detected possible seismic disturbances in the planet's crust.

Arecibo 12.6-cm-wavelength radar image (1988), in the Mercator projection, of a tract bounded by 320° and 20° in longitude and by –10° and –45° in latitude. The diffuse bright spots aligned along a line more-or-less parallel to the left edge of the image are the volcanoes Ushas, Innini, and Hathor Mons. The three craters in a dark area left of center and arranged in an inverted L-shape constitute the "crater farm" imaged by Magellan early in the mission (Fig. 4.36). Radar-bright Alpha Regio occupies the right half of the image, along with several coronae. The ring-like structure to the south-southwest of Alpha Regio is Eve, a 300-km corona centered at (–32.0°, 359.0°). The image, with resolution of about 1.5 km, represents the very best snapshot of Venus that Earth-based radars have produced to date. [D. B. Campbell, Cornell University and Arecibo Observatory.] 1.7

━━━━━━━━ 600 km

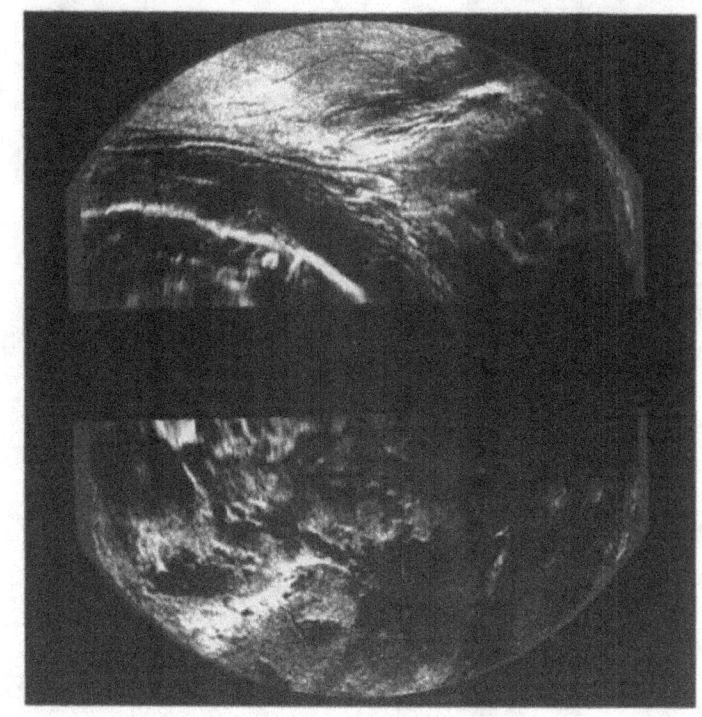

Goldstone 12.9-cm-wavelength radar image (1988) of a segment of the Heng-O structure. The image is centered at (4.2°, 357.2°); the diameter of the structure is 1060 km, and its center is at 2.0°, 355.0°. Image resolution is approximately 1 km. Classification of the structure as a corona, based on this image and on earlier Goldstone images, was confirmed by Magellan data (Fig. 5.16). *[R. F. Jurgens, Jet Propulsion Laboratory.]* 1.8

150 km

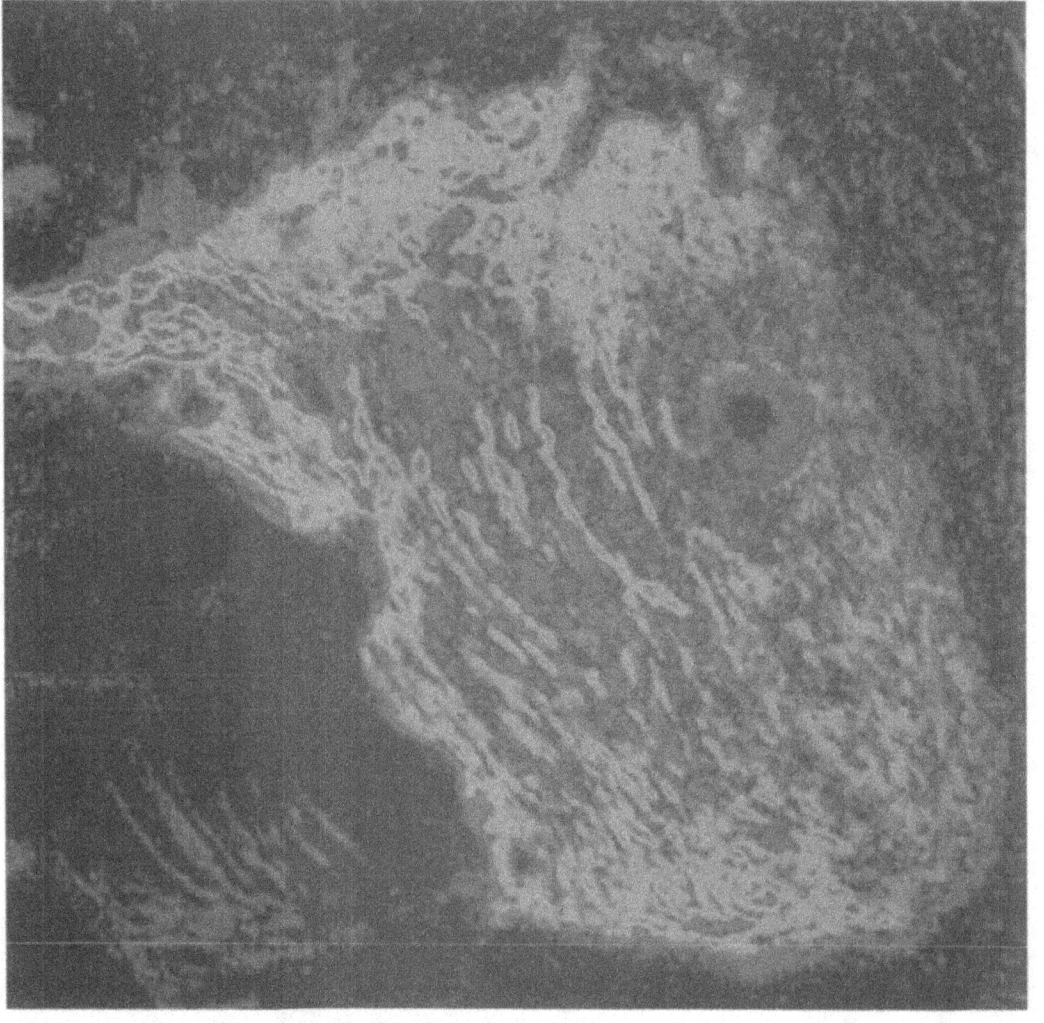

The extraordinary capability of Earth-based radars is demonstrated in this Arecibo 12.6-cm-wavelength image (1983) of Maxwell Montes, a massive mountain range in the eastern part of Ishtar Terra and another of the bright areas noticed in the early radar images of Venus. Resolution of the image is 3 km. The banded appearance of the Montes suggested, even before the arrival of Magellan at Venus, a complex, tectonically reworked terrain. The circular feature near the right edge of the Montes is the crater Cleopatra (diameter 105 km; 65.9°, 7.0°; Fig. 4.20), initially thought to be a volcanic depression. *[D. B. Campbell, Cornell University and Arecibo Observatory.]* 1.9

VENUS in 1934.
South.

North.

March 31.
9¾ hrs.
× 210.

April 20.
8¾ hrs.
× 210.

May 2.
9¾ hrs.
× 210.

Drawings by R. Barker of the aspects of Venus as observed during the 1934 apparition. The dark streaks are supposedly identical to those mapped by Lowell. In Barker's words, "the streaks on Venus are always diffused, and (in broad sunshine), appear as smoke-coloured streaks on the pale disc, and, although more difficult to grasp than the best known martian 'canals,' are unmistakable if viewed with a really good telescope." Further, "these streaks, seen through a cloudless venusian atmosphere, prompted the remark from Lowell that 'we are looking down on a bare, desert-like surface.'" While Barker's scientific peers more or less accepted the reality of the elusive, ill-defined, dusky lineations on the venusian crescent, they differed on the degree of the lineations' permanence and disputed the suggestion that the planet's surface might be visible. The numbers attached to the individual streaks in the sketches refer to Lowell's nomenclature. 1.10

In 1983, orbiters Venera 15 and 16 carried one step further the work begun by the Pioneer Venus Orbiter by acquiring higher resolution radar imaging and altimetry data over the northern latitudes of the planet. The imaging data had a resolution of 1–2 kilometers, comparable to the best data obtained by Earth-based radars; altimetry data resolution was a factor of four better than that of Pioneer. In a particularly imaginative adventure, in 1985, each of the Vega (Venus–Halley) 1 and 2 landers released a helium-filled balloon about 50 kilometers above the planet's surface, approximately in the middle of the venusian three-deck cloud system. The free-drifting balloons were intended to provide information on the dynamics of the atmosphere's most active section.

An image of of the day side of Venus, taken February 14, 1990, by the solid-state imaging system on the Galileo spacecraft. The image, taken through a violet filter at a distance of 2,700,000 km, indicates existence of convective activity in the topmost layers of the sulphuric-acid clouds enveloping Venus. To reach Jupiter, its ultimate destination, Galileo had to be launched on a trajectory involving a Venus flyby. 1.11

The formidable accomplishments of Earth-based
radars and of the Mariner, Venera, Pioneer, and Vega
series of missions paved the way for the success of
the Magellan mission. By achieving almost complete
global coverage in imaging, altimetry, and gravity
and radiometry mapping, Magellan has established
a benchmark in the history of planetary exploration.
The subsequent chapters present a survey of Magellan's contributions to an emerging understanding of
the geological evolution of Earth's sister planet.

Magellan's accomplishments have established a benchmark in the history of planetary exploration

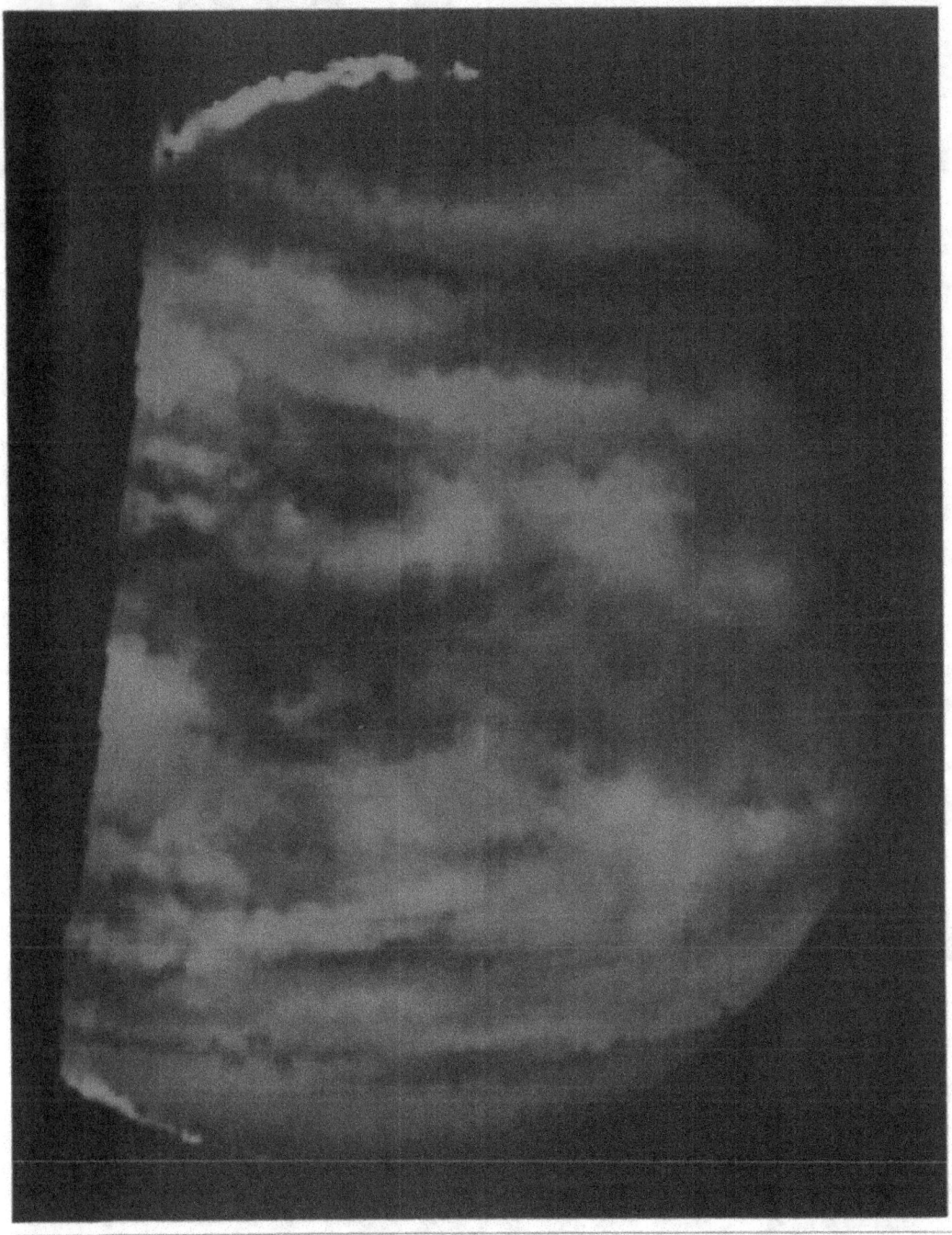

A map of the cloudy night side
of Venus, constructed from
Galileo near-infrared mapping
spectrometer 2.3-micron data —
acquired on February 10, 1990, at
a range of about 100,000 km. The
red hues represent the heat from the
lower atmosphere, radiating through
discontinuities in the the mid-level
(about 50 km above the planet's surface) cloud system. Wind speeds at
this level are estimated to reach about
250 km/hr. **1.12**

Mission to Venus

Radar was initially conceived as a device for determining the position and motion of discrete targets. Advances in electronics technology and signal processing permitted the scope of radar to also include imaging of distributed targets. Between 1960 and 1990, a number of airborne radar systems were built with the purpose of implementing the new concept and for imaging areas of geological interest. The first radar system intended for acquiring images of planetary landscapes, designed and built at the Jet Propulsion Laboratory (JPL), was incorporated into the Apollo 17 Command Module and operated during two full orbits of the Moon in 1972. The success of the experiment confirmed the feasibility of radar-imaging missions to other planets — Venus in particular.

View of the Magellan spacecraft as it leaves the space shuttle Atlantis. The high-gain antenna (HGA) dish (top) carried out radar mapping and communications with Earth. The horn at the left side of the HGA is the radar altimeter antenna. Below the HGA is the forward equipment module, which houses radar sensor electronics, batteries, and reaction wheels. The polygonal body attached to the equipment module is the bus, which includes the command and data system, attitude-control propellant, tape recorders, and other electronic subsystems. Spacecraft power is supplied by two solar panels, which are still folded in this view. The cylindrical structure houses the inertial upper stage (IUS) rocket engine. Together, the spacecraft and IUS engine measured 11.6 m in length; their combined mass was 18,233 kg. **2.1**

Shortly before the 1972 flight of Apollo 17, the Mariner 9 spacecraft completed an extraordinarily successful mapping mission to Mars. Conceived with the objective of optically mapping 70 percent of Mars' surface over a 90-day period, Mariner 9 instead remained in operation for 349 days, mapping almost all of the martian surface at 1–2-kilometer resolution as well as 2 percent of the surface at 100–300-meter resolution.

A mission to Venus, named the Venus Orbiting Imaging Radar (VOIR), was to accomplish a Venus global-mapping feat similar to that of Mariner 9 at Mars — map at least 70 percent of Venus' surface in synthetic aperture radar (SAR) mode at a resolution better than about 400 meters. The radar sensor was also to collect radio-emission and altimetry data over the imaged portions of Venus' surface. After suffering through the customary vicissitudes, VOIR was cancelled in 1982. In 1983, it was replaced with a more focused, simpler mission, provisionally named

the Venus Radar Mapper (VRM). Nonradar experiments were removed from the projected payload, but the basic science objectives of VOIR — investigation of the geological history of the surface and the geophysical state of the interior of Venus — were retained. Hughes Aircraft Company was chosen as the prime contractor for the radar system, the Martin Marietta Astronautics Group was given responsibility for the spacecraft, and JPL was entrusted with overall management of the mission. In 1986, while still in the developmental stage, the mission was given a definitive name in honor of Ferdinand Magellan (Fernao de Magalaes, 1480?-1521) — the first seafarer to attempt the circumnavigation of Earth.

The Magellan spacecraft was launched aboard the space shuttle Atlantis (STS-30) on May 4, 1989. In Earth orbit, the inertial upper stage (IUS) placed the spacecraft into a Venus transfer orbit. Insertion into orbit around Venus was accomplished by firing a solid rocket motor (later jettisoned) located between the solar panels. Attitude control and orbital maneuvers at Venus used hydrazine-fueled rocket-engine modules mounted on spider-like outriggers. In order to contain costs, only about 30 percent of the spacecraft was designed specifically for the Magellan mission; the rest was assembled mostly from spare Voyager and Galileo mission components.

On August 10, 1990, Magellan arrived at Venus. A solid rocket motor burn slowed the spacecraft and placed it in a near-polar orbit with periapsis latitude of 9.5 degrees, periapsis altitude of 289 kilometers (above the 6051-kilometer radius), apoapsis altitude of 8030 kilometers, and a period of 3.26 hours. The data-collection period of about 37 minutes duration per orbit would start near the planet's north pole (with the spacecraft at an altitude well above 2000 kilometers), continue southward through the periapsis, and finish by mapping at an altitude well above 2000 kilometers. The incidence angle would vary from about 15 degrees at both extremes of a

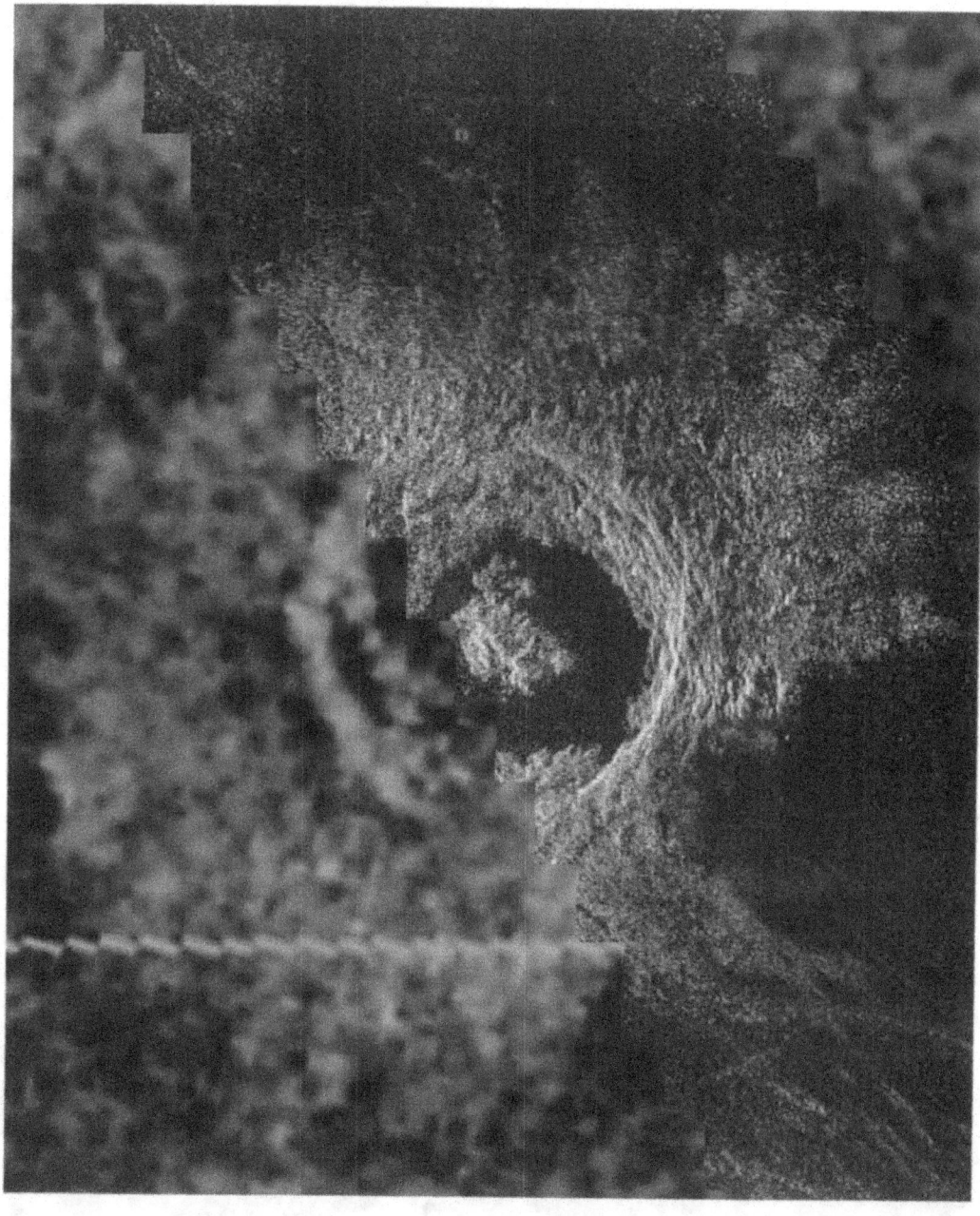

An image strip acquired during the first test of the Magellan radar was laid over Venera 15/16 data to produce this composite image of the crater Golubkina (diameter 30.5 km; 60.3°, 286.6°. It is evident that Magellan images permit geological investigations at a level of detail that was previously inaccessible. **2.2**

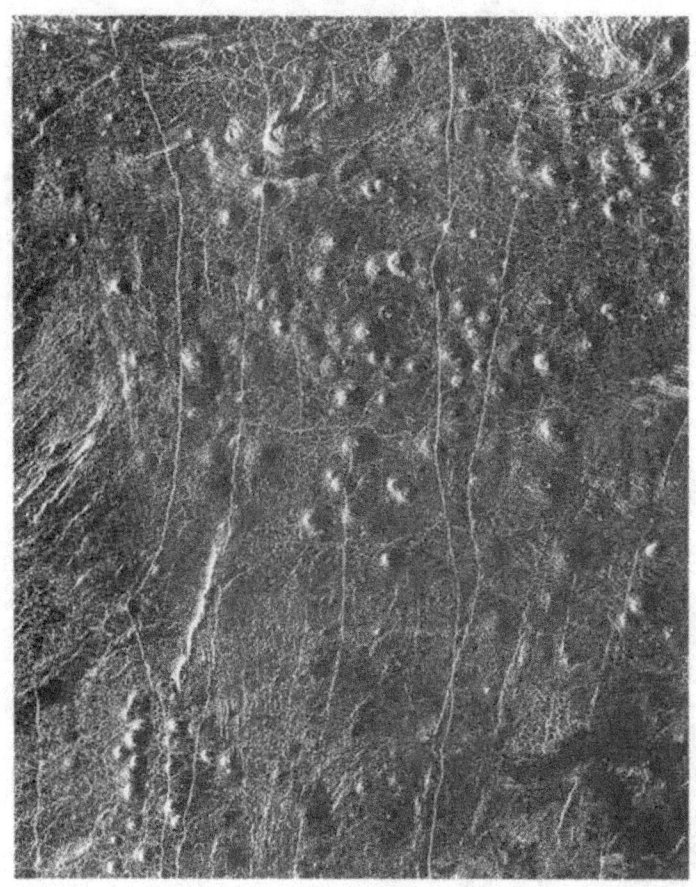

Radar images of an identical target area in Venus' Tethus Regio. The Magellan image is at left; the Venera 15/16 image is below. Hillocks, ranging in diameter from about 2 km to about 12 km and barely discernible in the Venera scene, have since been identified as volcanic shields. The contrast in the information content of the two images is due to the superior resolution of Magellan data. The resolution in azimuth (along track) is about 120 m; the resolution in range (across track) varies from about 120 m at periapsis to about 300 m at both ends of a data-collection run. Brightness variations in a radar image are generated mostly by variations in the microtopography of the imaged surface. In simplest terms, a target rough at the radar wavelength would generate a strong echo and would appear bright, while a smooth target would appear dark. The wavelength of Magellan radar was 12.6 cm; thus, the presence of blocks in the range of sizes from several centimeters to several meters within the area illuminated by the radar beam would result in a bright image. Because radar operates at a single frequency, radar images are perforce monochromatic. Polychromatic rendition of radar images involves an arbitrary assignment of color hues. **2.3**

data-collection run to about 47 degrees at periapsis. Daily, the spacecraft acquired 7.3 orbits of image data, with each orbit covering a strip 20–25 kilometers wide and 17,000 kilometers long. Between two successive orbits, due to Venus' rotation, the imaged strip would shift 20.54 kilometers at the equator. The overlapping strips were mosaicked to yield synoptic views of large tracts of the venusian surface. Coverage of the entire planet would require about 1800 strips.

The Magellan radar was designed to operate in the so-called "burst" mode, in which SAR imaging, altimetry, and radiometry share a single time slot — the burst period. Within each period, the interleaved 5-kilohertz SAR transmit and echo pulses were handled first. The elliptical trajectory required a continual repositioning of the echo window. Consequently, radar operating parameters had to be adjusted about 4000 times during each data collection run (each orbit), a sharp departure from practices typical for

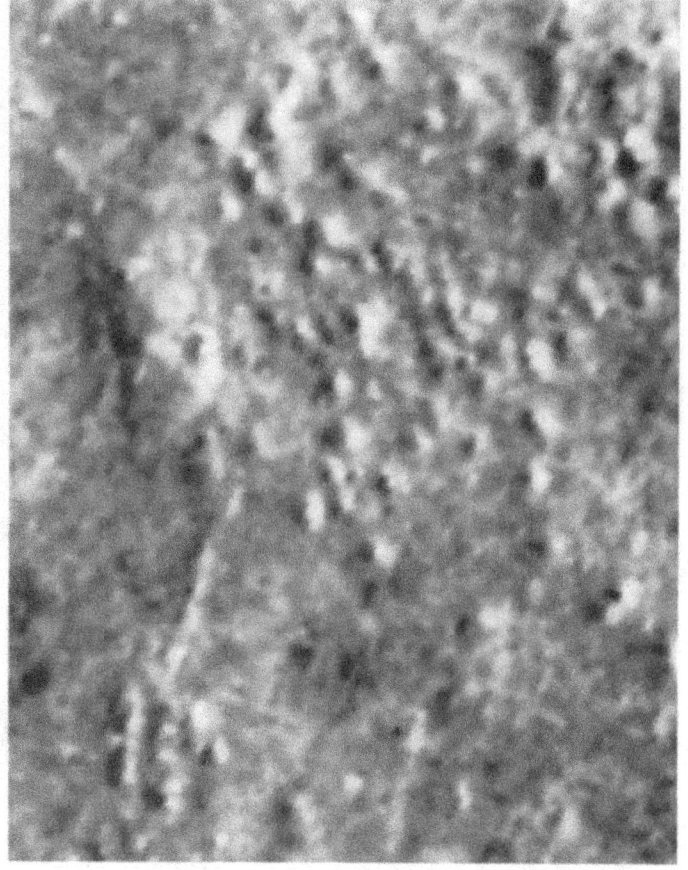

imaging radars moving parallel to the target surface. The commands required for computing the updates were stored on board the spacecraft and updated as necessary. After the last of the SAR echoes was received, the radar would transmit and receive a brief round of 15-kilohertz altimetry echoes. In the last operation of the burst period, the radar would act as a passive radiometer, receiving the microwave thermal energy from the sources both within view of the high-gain antenna as well as those along the signal path within the radar system. The sequence of operations, from SAR imaging to radiometry, would be repeated in the subsequent burst period. The length of a single burst period varied from about 250 milliseconds to about 1200 milliseconds; the number of

By the end of the first mapping cycle, Magellan had mapped about 84 percent of Venus' surface

M osaic of Magellan image strips covering the neighborhood of the Venera 8 landing site (marked as BEHEPA 8) in Navka Planitia (−10.0°, 335°). The Venera 8 lander transmitted from Venus' surface for 50 minutes. According to the results of gamma-ray spectrometer analysis, the rocks below the lander resemble alkali basalts on Earth. Magellan images allow linking of information about surface composition obtained by Venera landers to venusian surface morphology. **2.4**

————————— 50 km

periods was such as to fit within the 37-minute data-collection window. All the data acquired during each orbit, a total of almost two billion bits, were recorded on board the spacecraft and transmitted during the same orbit to the receiving stations of the Deep Space Network

The first images of Venus were collected on August 16, 1990. Routine mapping operations, occasionally marred by minor — albeit onerous — technical problems, began on September 15, 1990. The first mapping cycle (Cycle 1), lasted 243 Earth days — the time it takes Venus to turn once on its axis beneath the plane of the spacecraft orbit. Cycle 1, which ended on May 15, 1991, was devoted to accomplishing the basic mission objectives: mapping about 84 percent of the planet.

With the successful completion of Cycle 1, during which Magellan performed "left-looking" imaging.

Juxtaposition of data from different sources illustrates the dramatic progress that has been achieved in the reconnaissance of Venus. Image strips acquired by Magellan during its first two orbits are superimposed here over Pioneer Venus altimetry data. Altimetry and imaging represent two distinct radar techniques, with imaging providing an inherently higher resolution. The image shown is centered at approximately (0.0°, 288.0°). **2.5**

15 km

Suite of images of the same target area, acquired during the first three Magellan mapping cycles, demonstrating the influence of viewing geometry on the visual appearance of a radar image. The spacecraft was looking to the left of the direction of motion when it produced the Cycle 1 (top) and Cycle 3 (bottom) images; it was looking to the right when the Cycle 2 (middle) image was produced. In the first case, the target was imaged from the west; in the second, from the east. Brightness differences between the Cycle 1 and Cycle 3 images on one hand, and the Cycle 2 image on the other, may be caused by reflections from stable but asymmetric sand or dust deposits — rather than by echoes from deposits susceptible to a wind-induced rearrangement. The imaged area is centered at (–47.5°, 226.0°). **2.6**

————————— 50 km

expanded data-acquisition objectives were specified for subsequent cycles. Cycle 2, lasting through January 15, 1992, was utilized to image Venus in "right-looking" mode and to fill the gaps in coverage that remained after the completion of the left-looking Cycle 1. Total coverage was brought to 92 percent. During Cycle 3, which was to conclude on September 14, 1992, 22 percent of Venus' surface was imaged in stereo and aggregate coverage grew to 98 percent. Mounting equipment problems led to the suspension of radar imaging on September 13, 1992. In the course of its radar operations, Magellan had generated a volume of data larger than the combined production of all preceding planetary missions.

The remainder of the Magellan mission was devoted to the collection of gravity data. Deviations from the radial distribution of mass in a planet create nonuniformities in the gravity field that perturb the motions of an orbiting spacecraft. By tracking the spacecraft, mapping the orbital disturbances, and

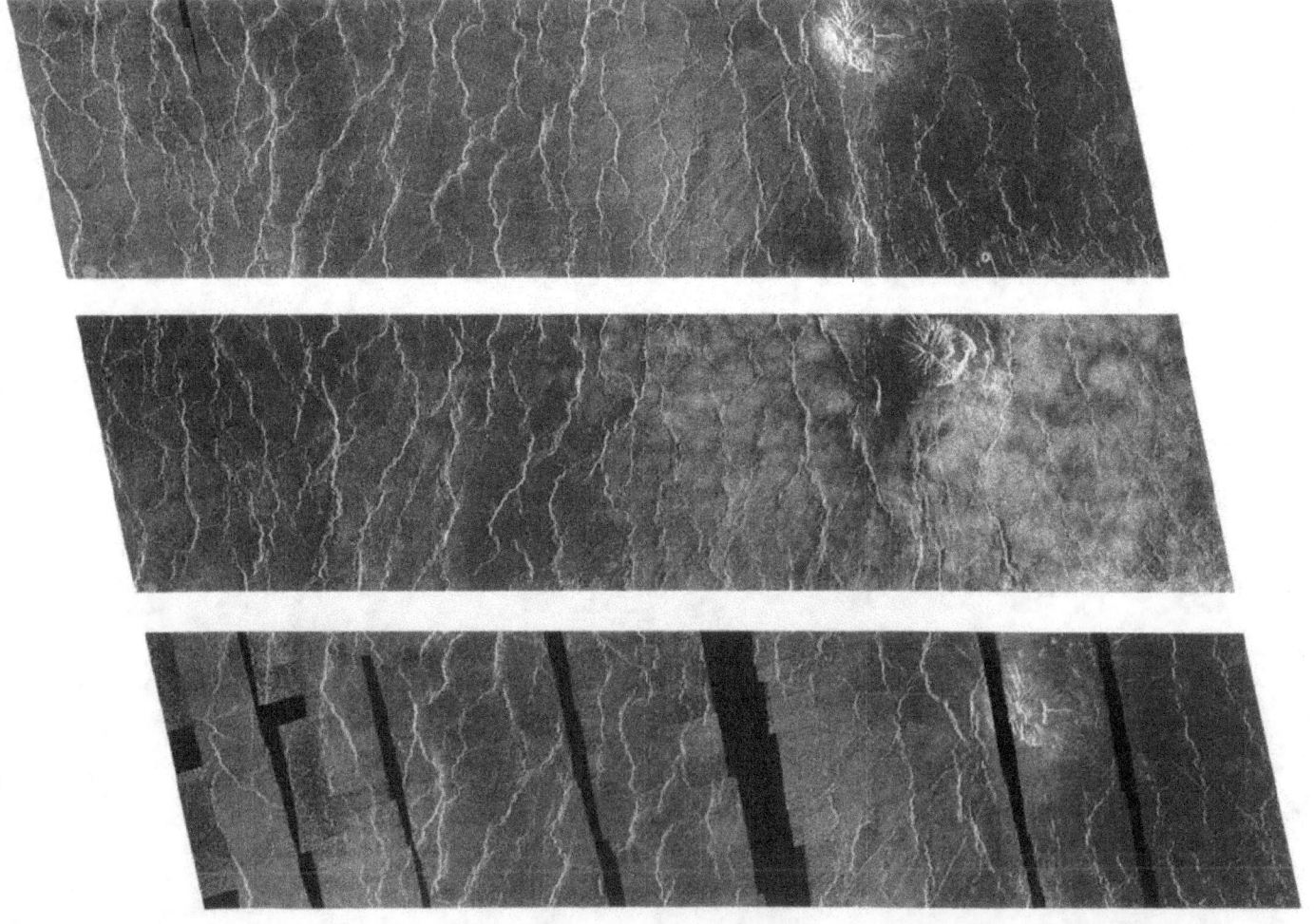

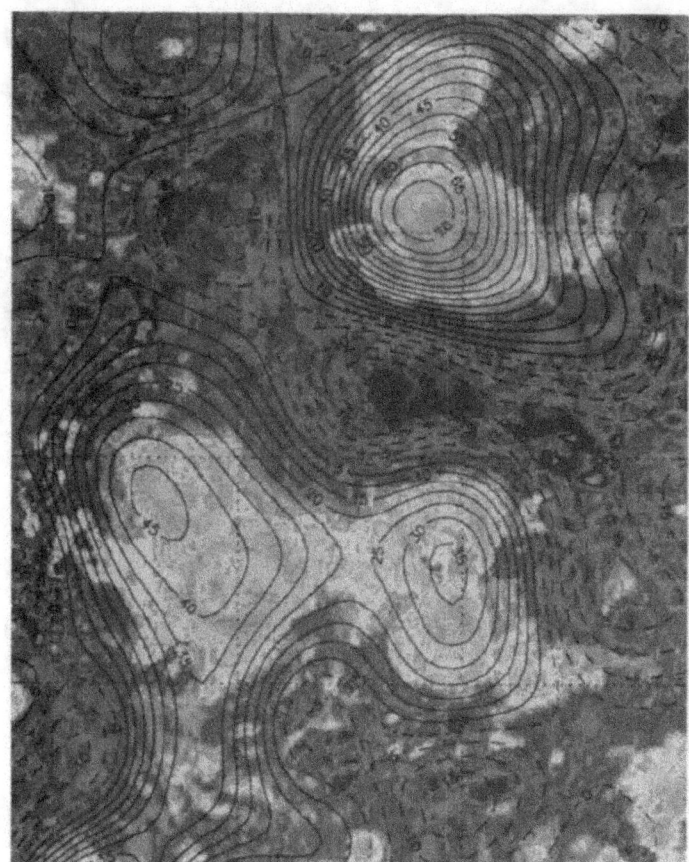

ontours of free-air gravity anomaly, in units of 10^{-5} m/sec^{-2}, extracted from Pioneer Venus (left) and Magellan (below) tracking data and superimposed on topography derived from Magellan altimetry data. By tracking at X-band (3 cm) — rather than S-band (13 cm) as in past missions — Magellan measured variations in spacecraft velocity to a higher accuracy than ever before possible. As a result, a noticeably closer agreement was achieved between the gravity anomaly and the elevation contours. [A. S. Konopliv and W. L. Sjogren, Jet Propulsion Laboratory.] 2.7

600 km

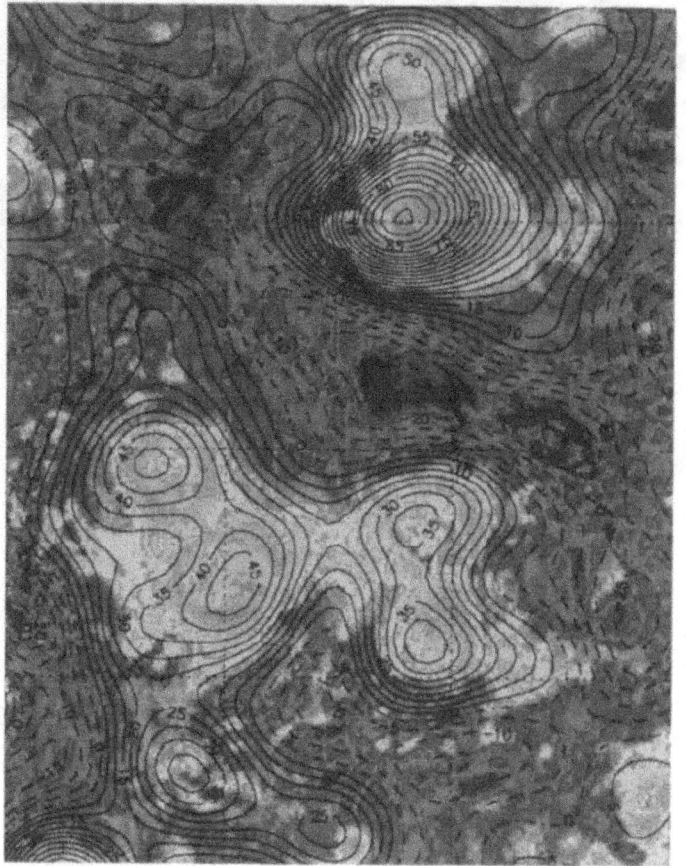

correlating them with the topography, inferences can be drawn about density variations in the planet's interior. Cycle 4 of the Magellan mission, which ended on May 24, 1993, was devoted to collecting gravity data over the entire planet.

Optimum gravity-data acquisition strategy calls for low-altitude, circular orbits. In the case of Venus, the minimum safe orbit altitude is about 200 kilometers above the surface. Cycles 5 and 6, running through October 31, 1994, were assigned to gravity-data collection, with the spacecraft moving in a low, circularized orbit. Magellan was brought to a near-circular orbit during a succession of aerobraking maneuvers that ran from the end of Cycle 4 through the beginning of August 1993. Aerobraking employed the drag of the planet's atmosphere to lower the apoapsis. To forestall the destruction of the spacecraft during this delicate phase, tracking and commanding had to be particularly precise.

The Surface

With the introduction of the telescope, Venus became the object of systematic optical observations — though their import had been much debated. Some astronomers believed they saw enduring dusky streaks through the supposedly transparent venusian atmosphere or through ephemeral gaps in the planet's opaque clouds (Fig. 1.3). Others claimed to have observed permanent luminous marks — glints of sunlight reflected from the cloud-piercing mountain tops; still others dismissed the very idea of anything permanent in the appearance of the planet's crescent. J. H. Schroeter, a respected observer and correspondent of W. Herschel, reported sightings between 1788 and 1790 of an "enlightened mountain" of extreme height, close to the terminator, in Venus' dark hemisphere. Although attacked harshly by Herschel and never confirmed, this observation is still occasionally cited.

A mosaic of Magellan radar strips projected onto Venus' sphere yielded this global view of the planet's surface, centered on the longitude of 180°. The bright belt meandering through the image, a continent-like highland named Aphrodite Terra, corresponds in area roughly to Africa. The prime meridian of the geodetic net passes through the north pole and through the central peak of crater Ariadne (43.95°, 0°). Because Venus rotates in a retrograde direction, longitude by convention increases to the east. (A grid-free, false-color version of the same image is featured on the front cover of this book.) 3.1

Schroeter's claim even found its way into the *Mariner–Venus (Mariner 2) Final Report*, where, beside a sketch of a grouping of grotesquely tall mountains, we read that "one scientist believed he identified a mountain peak [on Venus] ... which he calculated as rising more than 43.45 kilometers (27 miles) above the general level of the surface." Observations of terminator irregularities were reported often enough to earn some credence, but the cause of these anomalies was most likely the uncertain nature of the planet's atmospheric veil.

Not surprisingly, the altimetry data gathered by Pioneer Venus, Venera 15/16, and Magellan radars provided no evidence for the existence of mountain crests tens of kilometers high. On the contrary — the global topography of Venus is quite tame. Of the 93 percent of Venus' surface mapped by Pioneer Venus, the total relief (from lowest to highest points) was found to be about 13 kilometers.

On Earth, by comparison, the total relief (from the bottom of oceanic trenches to the tops of Himalayan peaks) is about 20 kilometers. According to the Pioneer altimetry data, approximately 51 percent of the surface of Venus lies within 500 meters of the mean radius of 6051.9 kilometers; only 5 percent of the surface is at elevations more than 2 kilometers above the mean radius. The Magellan altimetry experiment confirmed the general lack of large-scale relief. In Magellan data, over 80 percent of the surface lies within 1 kilometer of the mean radius. The highest elevations on Venus are reached within the mountain ranges that girdle Lakshmi Planum: Maxwell Montes (11 kilometers above the datum), Akna Montes (7 kilometers), and Freyja Montes (7 kilometers). Although Venus' topography lacks dramatic excesses, the Magellan altimeter did identify areas of exceptional steepness. Parts of the southwestern wall of Maxwell Montes may be inclined by as much as 45 degrees.

Slopes of about 30 degrees have been measured in Danu Montes and along the chasmata walls east of Thetis Regio.

Based on Pioneer Venus altimetry data, the surface of Venus is divided into three topographic provinces: lowlands, rolling plains, and highlands.

Magellan altimetry data support this division. The major highland provinces include Aphrodite Terra, Ishtar Terra, Lada Terra, and the aggregate of Beta, Phoebe, and Themis Regiones. Alpha, Bell, Eistla, and Tellus Regiones comprise an assemblage of lesser highlands.

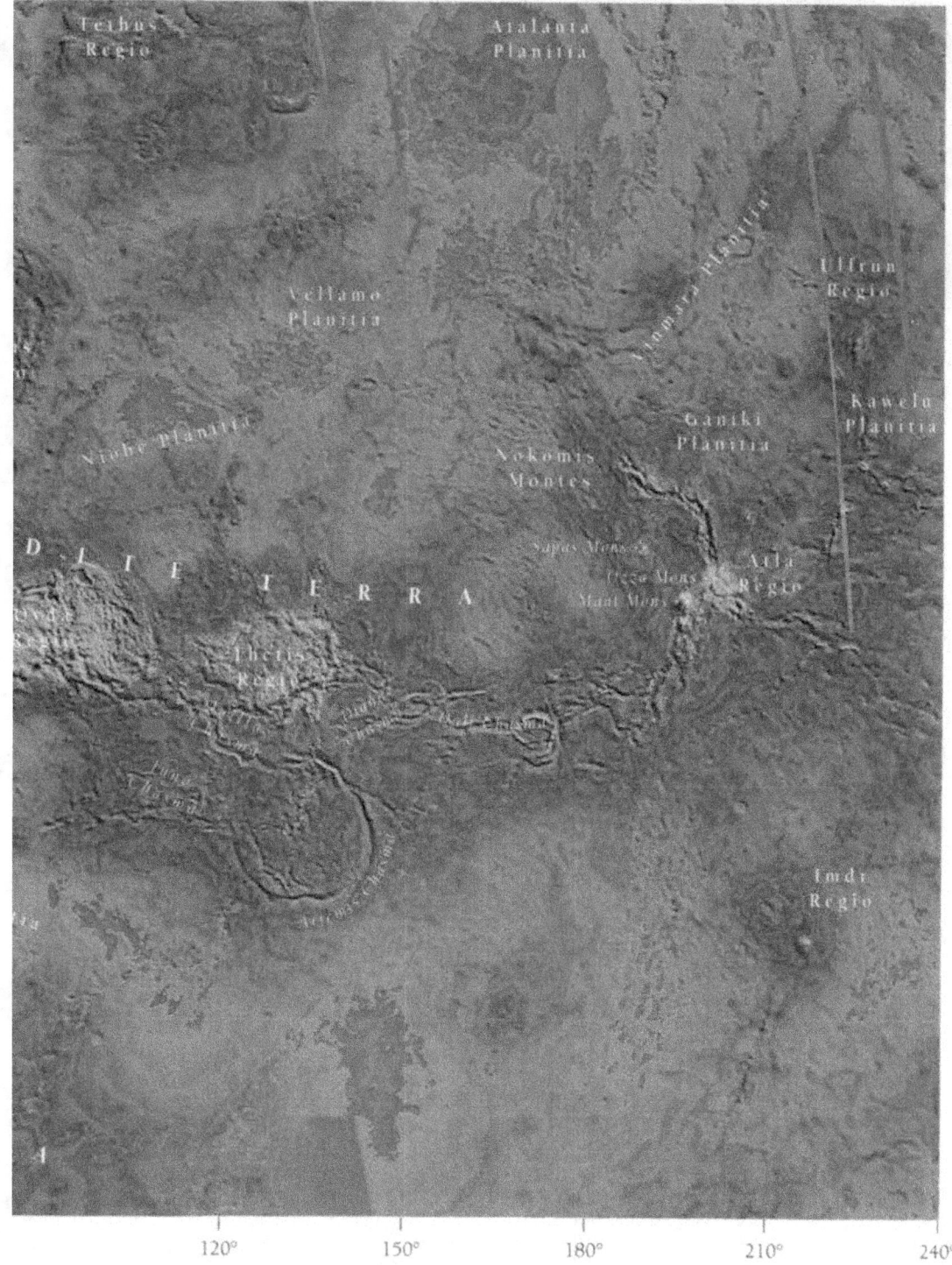

Teihus Regio

Atalanta Planitia

Vellamo Planitia

Ulfrun Regio

Niobe Planitia

Kawelu Planitia

Ganiki Planitia

Nokomis Montes

Sapas Mons

Atla Regio

Ozza Mons

Maat Mons

D I E   T E R R A

Theris Regio

Imdr Regio

| 120° | 150° | 180° | 210° | 240° |

Topographic map of Venus, compiled from Magellan altimetry data. Red hues represent the highest elevations; blue hues represent the lowest elevations. The Magellan altimeter mapped Venus over latitudes from 85° in the north to –85° in the south, extending the Pioneer Venus coverage — which was restricted to latitude from 74° to –63°. The vertical resolution of the Magellan altimeter was about 80 m; horizontal resolution was about 10 km. The overall flatness of the venusian surface, first noted in Pioneer Venus data, is statistically characterized by a unimodal distribution of elevations. In contrast, elevations on Earth are distributed bimodally. While terrestrial continents very roughly correspond to venusian continents, oceanic basins on Earth have no equivalents on Venus. **3.2**

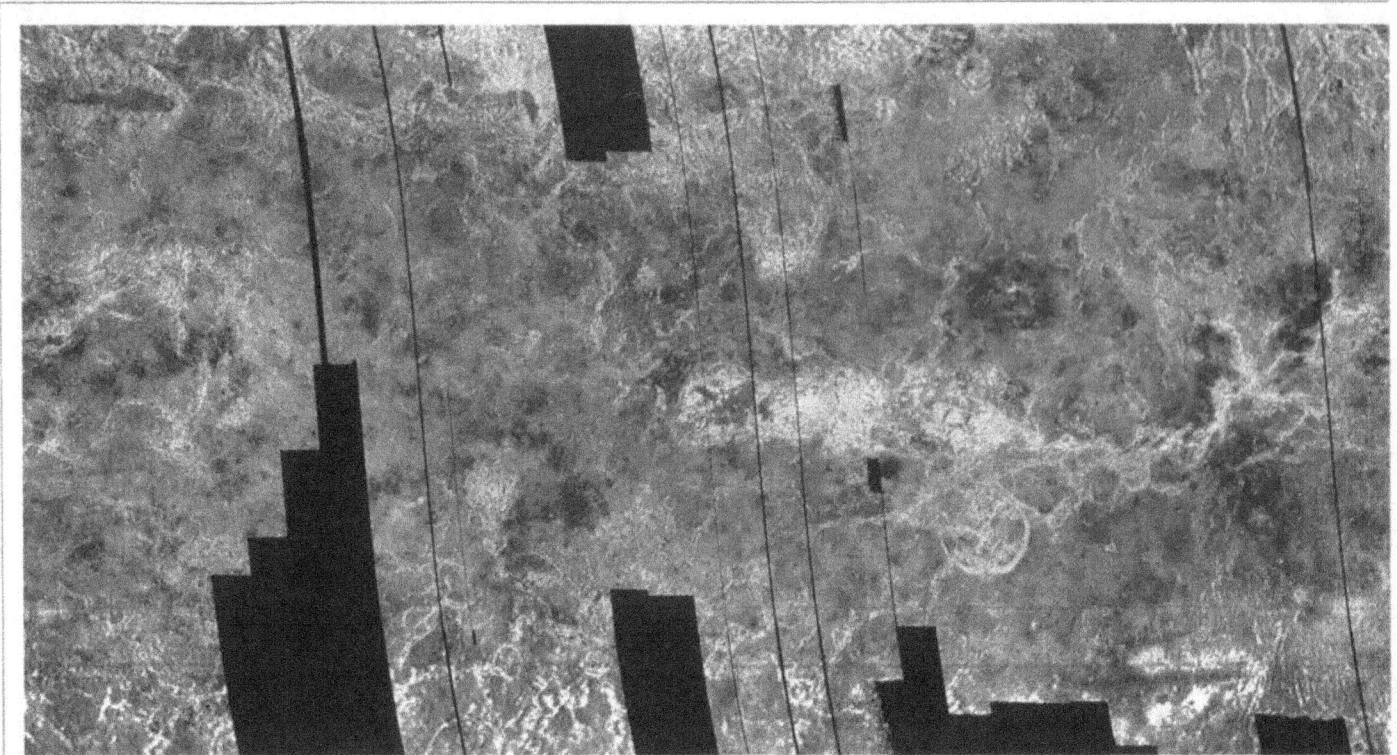

Radar echoes contain information on the small-scale geometry of planetary surfaces. Radar altimeters illuminate their targets from almost directly overhead. As a result, most of the energy scattered back toward the radar comes from terrain elements that are either horizontal or slightly tilted with respect to the horizontal. Reflection from such elements is mirror-like, or "specular." In the case of Venus-orbiting radars, elements producing specular or near-specular reflection may range in size from fractions of meters to many hundreds of meters. A statistical description of the surface, useful in understanding the radar echo, is its "root-mean-square" (RMS) slope or roughness. The closer the RMS slope is to 0°, the closer the target resembles a polished sphere.

In the simplest altimeter data reduction scheme, the return time of each radar echo is measured. The return time, linked with knowledge of the spacecraft position at the given instant, allows determination of the planet's radius for each measurement. But echoes return to the radar from surface points other than the nadir. In fact, for surfaces with large RMS slopes, the nadir echo may be a small part (albeit usually the first) of the total echo. In a more refined reduction scheme, the recorded echo-delay functions are matched to precomputed, model-derived functions. In this manner, estimates of planetary radius (Fig. 3.2), RMS slope (above), and reflectivity (Fig. 3.4) can be obtained simultaneously. Early Earth-based radar observations determined that Venus is smoother than both the Moon and Mars. The Pioneer Venus and Magellan roughness maps confirm that finding. Of the three principal physiographic provinces on Venus, the lowlands are on average the smoothest. The rough highland area of the rolling plains contains Alpha Regio (−25°, 5°), the first feature on Venus recognized by radar from Earth (1965). **3.3**

Reflectivity of Venus' surface, extracted from Magellan altimetry data. Radar reflectivity is a measure of the bulk permittivity (dielectric constant) of the reflecting surface. Since permittivities of common rocks are well-known, radar-determined reflectivity gives a fair assessment of the physical constitution (composition and density) of the uppermost layer of the planetary surface. On the average, Venus is about twice as reflective at centimeter-wavelengths as the Moon and Mars. Permittivity of compact, basaltic rocks best fits most of the Venus reflectivity data. Significant deviations from the global average were detected for elevated regions, in accordance with earlier findings from Arecibo, Pioneer Venus, and Venera 15/16. **3.4**

The surface of Venus is smoother than that of either the Moon or Mars

Root-mean-square (RMS) surface slopes, derived from Magellan altimetry data by direct inversion (rather than model-matching as in Fig. 3.3). Purple and blue hues in the map represent a preponderance of surfaces that are almost horizontal; yellows correspond to surfaces with the largest mapped RMS tilt (about 6°). The roughest surfaces are found in the area east of Beta Regio and in a number of locations within the Aphrodite Terra belt. *[G. L. Tyler, R. A. Simpson, M. J. Maurer, and E. Holmann, Stanford University.]* **3.5**

Total topographic relief on Venus is about 13 kilometers, compared to about 20 kilometers on Earth

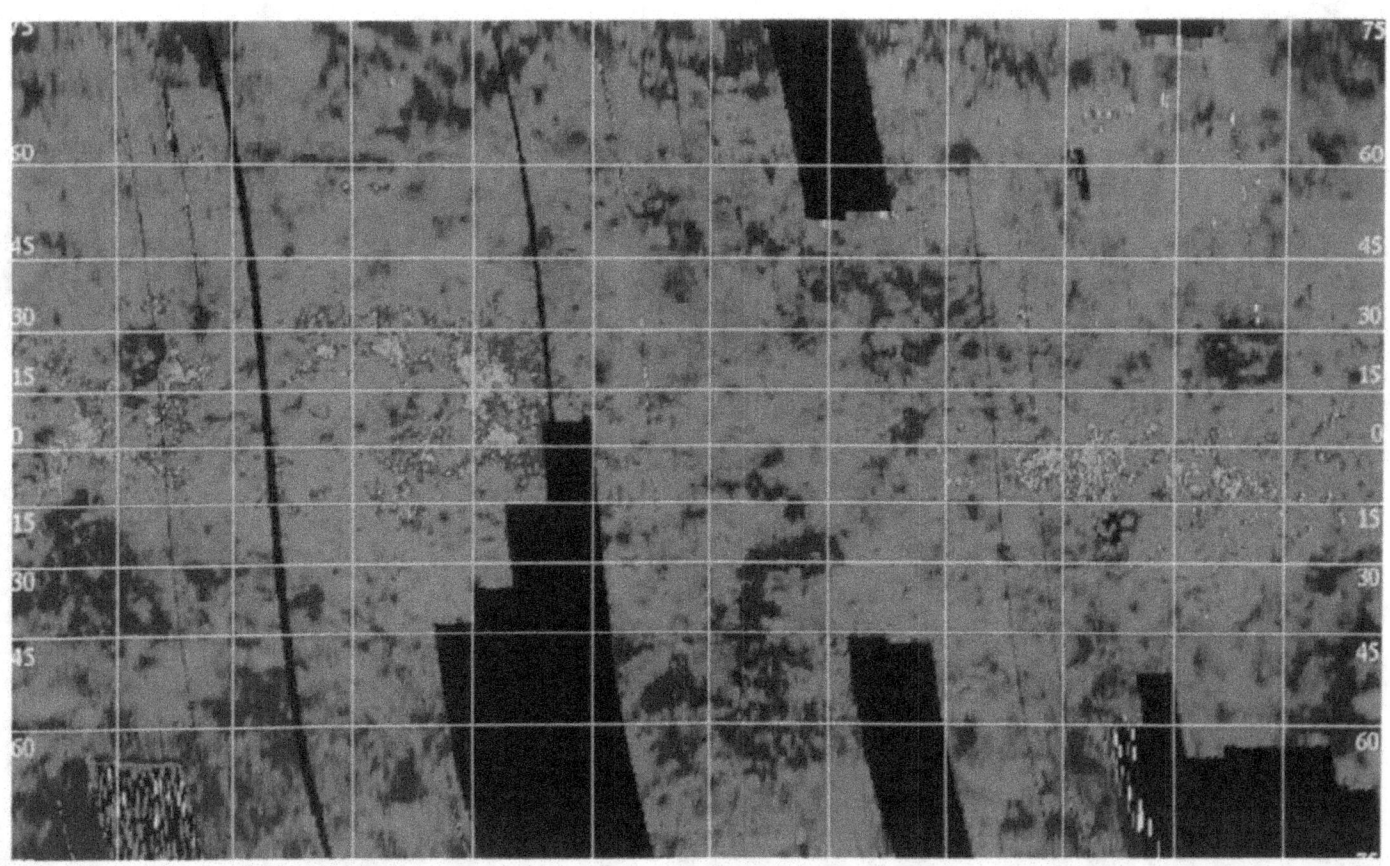

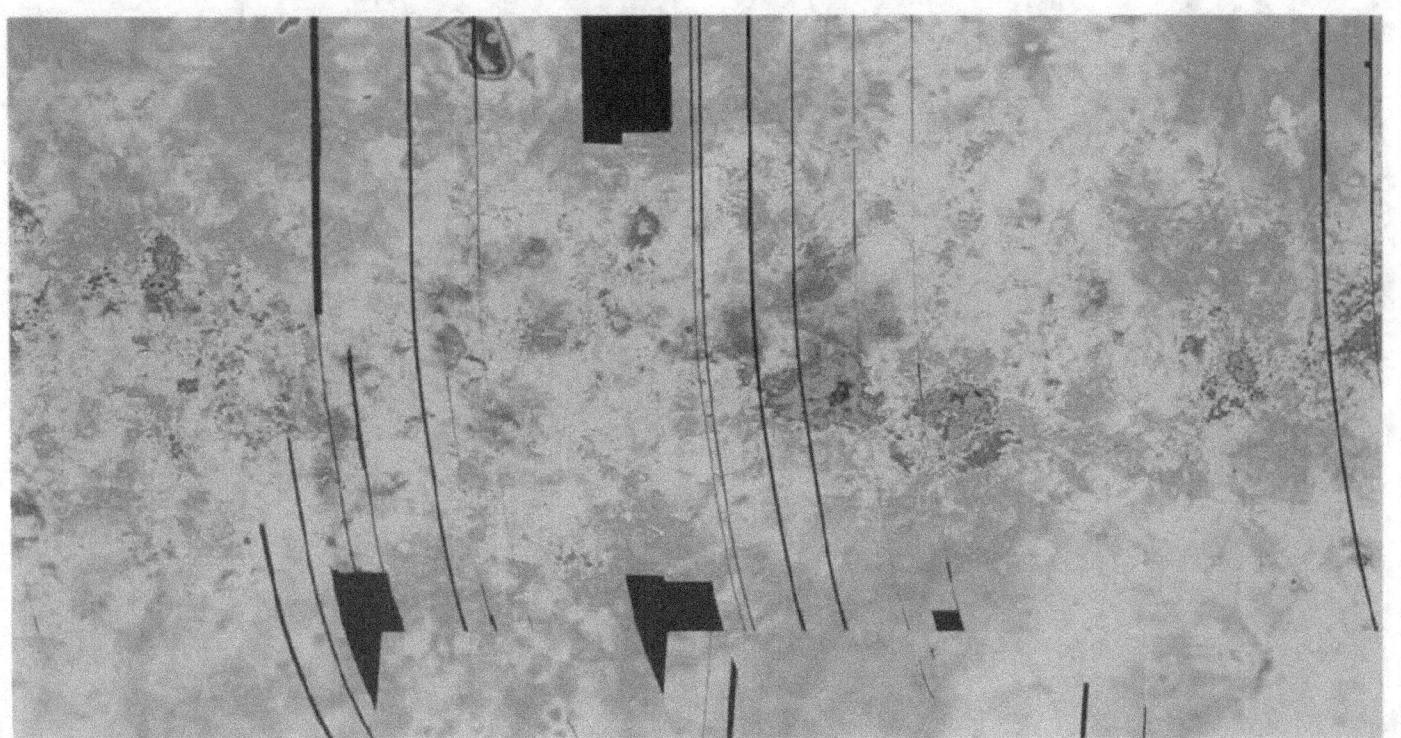

Magellan's radar, when operated in its passive, receive-only mode, was used to capture the energy radiated by Venus at the wavelength of 12.6 cm. Microwave thermal emission data may be used to estimate permittivity (dielectric constant) of a surface if its temperature is known. By combining emissivities of individual footprints (parcels of ground intersected by the 2-degree-wide Magellan high-gain antenna beam), a global emissivity map of Venus was compiled. The white and red hues in the map correspond to the highest emissivities (about 0.95 and 0.90, respectively), and the blue hues to the lowest (about 0.3). In simplest terms, a low-permittivity object is a poor reflector but a good emitter, and vice versa. The mean emissivity of the venusian surface mapped during Cycle 1 was found to be 0.845. An emissivity of comparable magnitude would characterize an object of permittivity 4.0–4.5. Dry, compact rocks of basaltic composition have permittivities that fall within that range. The reflectivity and emissivity data suggest that Venus' average surface composition consists of rocks that are analogs of terrestrial basalts. **3.6**

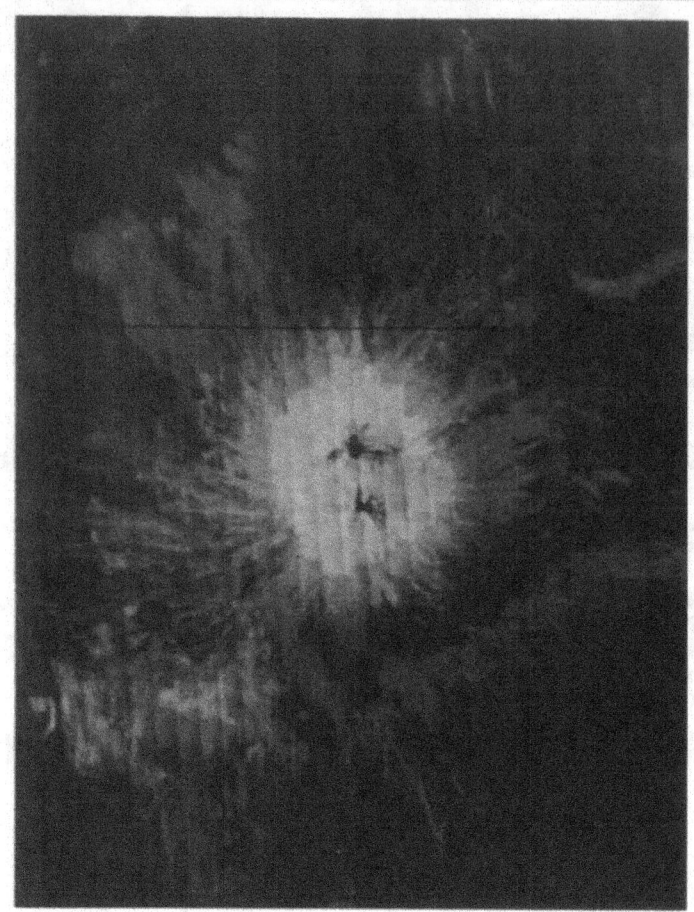

A degree of correlation between elevation and emissivity is one of the startling discoveries made in the course of radar exploration of Venus. Magellan topography (above left) and emissivity (above right) maps of the volcano Sapas Mons (8.5°, 188.3°) demonstrate this correlation. In general, the ordinary, high emissivity (red hues in the emissivity map) rocks are found at low elevations, while the low-emissivity (blue hues), high-reflectivity rocks are seen at higher elevations. The low-emissivity rocks at high elevations may signify the presence of rather uncommon minerals. Alternately, the correlation between low emissivity and high reflectivity at high venusian elevations may be due to the action of some as yet poorly understood mechanism of radar-echo generation. *[G. H. Pettengill, P. G. Ford, and R. J. Wilt, Massachusetts Institute of Technology.]* **3.7**

══════ 150 km

A recibo and Venera 15/16 data established volcanism and tectonism as the dominant geological processes that shaped the surface of Venus. In Magellan images, the products of the action of these processes are often expressed through patterns and textures of extraordinary beauty. The section of Atla Regio (–9.0°, 199.0°) exhibited (right) could be mistaken for a sheet from an avant-garde artist's sketchpad or for a naturalist's catalog of ocean-floor fauna. Yet this is geology in its purest form that is on display. In general, circular and oval configurations seen in Magellan images are the surface manifestations of volcanism, while bright striations and semilinear designs are the toil of tectonism. In some instances, oval-shaped landforms were created by the planet's collision with bodies in orbit about the Sun. The incidence and distribution of impact landforms provide the temporal framework for the reconstruction of a planet's geological history. Our survey of Magellan results thus begins in Chapter 4 with a discussion of venusian impact landforms. Images of volcanic features are shown in Chapters 5 and 6; images attesting to the possible presence of wind erosion on Venus are introduced in Chapter 7; and evidence for the tectonic reworking of large parcels of the venusian surface is reviewed in Chapter 8. **3.8**

══════ 30 km

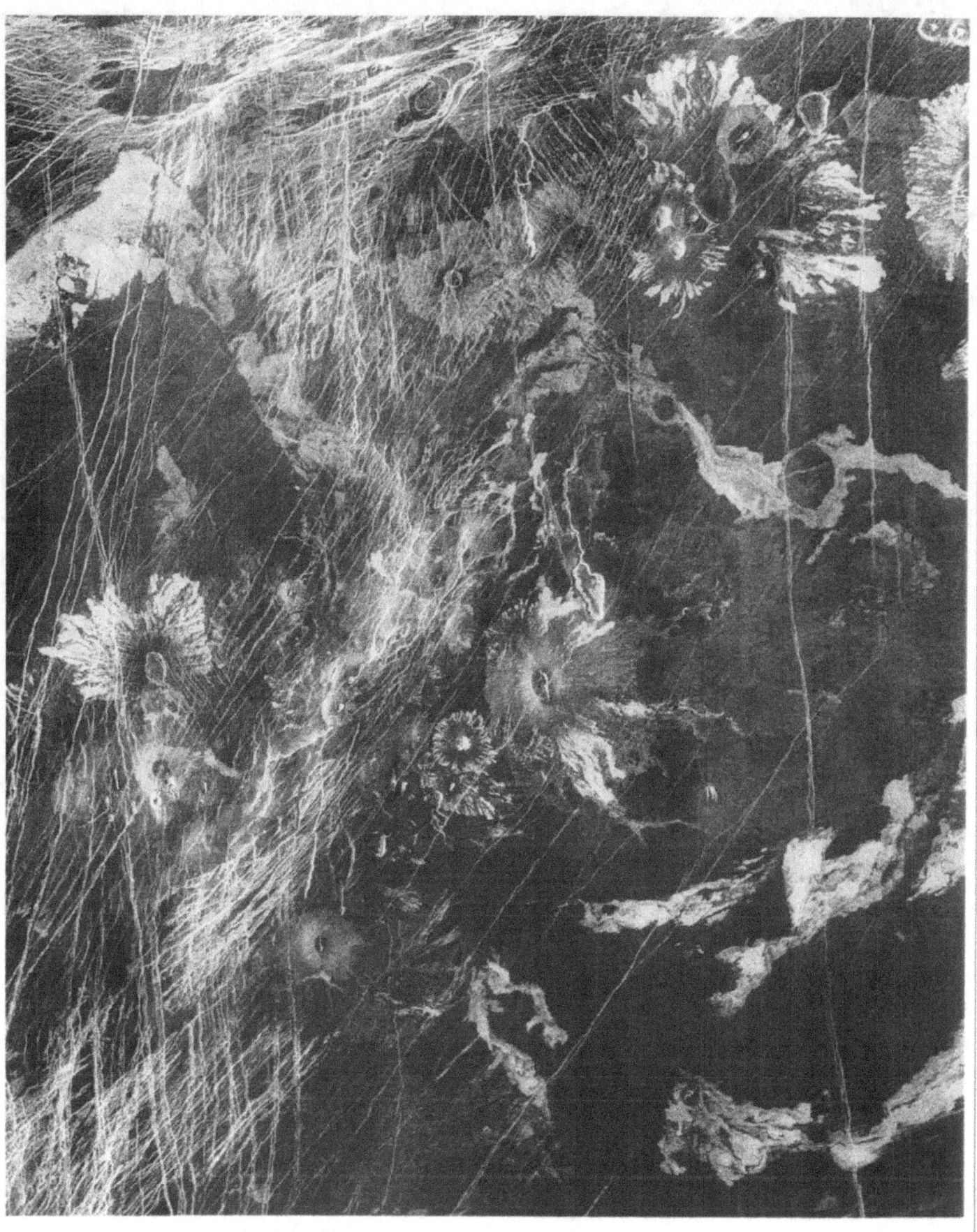

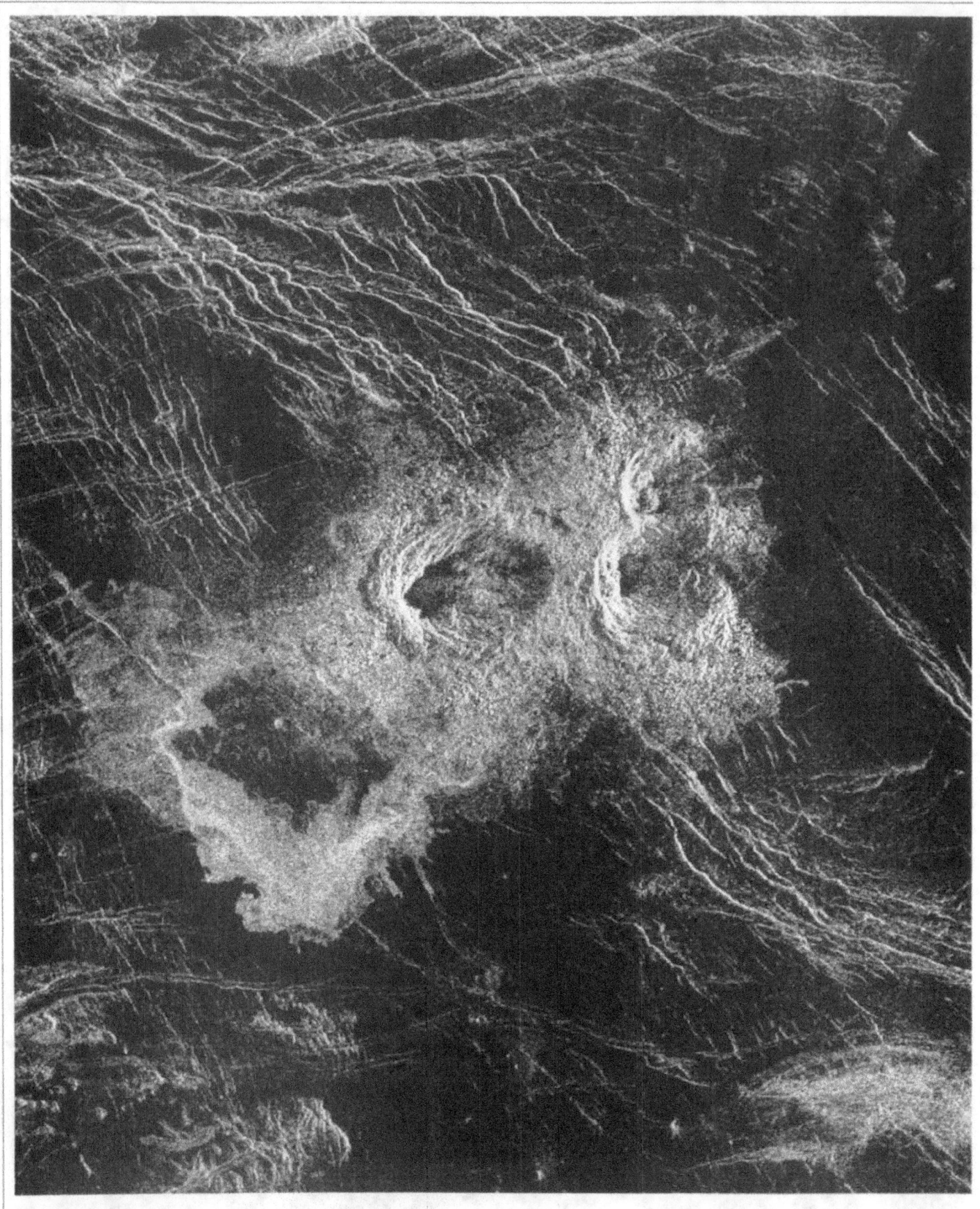

Impact Craters

Chapter 1 alluded to the discoveries Galileo made through his use of the newly invented telescope. Equipped with this novel tool, Galileo studied the Moon before he observed Venus. To his astonishment, Earth's companion did not seem to be a smooth crystalline sphere, as the belief of a "large school of philosophers" then held, but an uneven body "full of hollows and protuberances." The hollows Galileo saw — and named "craters" — were landscape forms unknown to him and his contemporaries.

Small projectiles hitting an airless planet or a planet with a thin atmosphere excavate craters referred to as "simple." The most notable characteristic of a simple crater is its uncomplicated (at the appropriate resolution), bowl-like shape. Simple craters of this type are rare in the Magellan data set. On Venus, small projectiles do not survive the atmospheric passage. Projectiles larger than those that undergo annihilation — but still not large enough to retain their structural integrity — are thought to generate irregular craters, multiple-floor craters, or crater clusters. The degree of crater irregularity is the function of the amount of dispersion of the fragmented meteoroid. Irregular craters of all descriptions constitute about 13 percent of venusian craters. Since crater clusters are formed by relatively small bolides, craters making up a cluster tend also to be rather small. The largest craters within the venusian crater clusters have diameters of about 40 km. Stein (diameter 15 km; −30.1°, 345.4°) is an example of a medium-size multiple-floor crater. 4.1

In ancient Rome, the term "crater" denoted a bowl for mixing wine with water. Ovid used the term when referring to volcanic vents — objects well known in the Mediterranean world. Galileo's use of the same term for the lunar hollows tacitly implied a volcanic connection. Even though mysterious — and long thought to be restricted to the Moon alone — craters have been found to be a common presence on the surfaces of all solid bodies observed thus far (except Jupiter's Io) in the solar system. The notion of the volcanic origin of most lunar craters was discarded during the course of the present century. We now view lunar craters as pits dug up by the Moon's collisions with chunks of matter in orbit around the Sun. Entities comprising the solar system — planets, asteroids, comets, etc. — very likely formed by incremental aggregation of smaller mass fragments upon the larger, more massive, gravitationally dominant nuclei. The process is believed to have been relatively rapid, lasting perhaps no longer than about 100 million years, and ending with the largest of the accretional nuclei — the planets —

scooping up most of the minor bodies and sundry unattached debris. The terminal phase of the growth of planets may have coincided with the period of profuse cratering, referred to as the late heavy bombardment. Vast tracts of the Moon, Mercury, and Mars bear witness to this stage of planetary development. On Earth, geological activity effectively destroyed the primordial craters, while some craters excavated since the conclusion of the era of heavy bombardment have survived. In the broadest sense, Earth's occasional brushes with stray meteoroids may be seen as a part of the drawn-out, closing act of the drama of planetary formation.

Venus, a terrestrial planet, was also expected to have its face marked by impact scars. Indeed, crater-like structures on Venus were identified in images produced by Earth-based radars several years before radar-equipped spacecraft reached the planet. The Venera 15 and 16 probes recognized almost 150 features of probable impact origin. Magellan, with its higher resolution and near-global coverage, tallied well over 900 likely impact craters. This number is low, considering the large surface area of Venus. Due to atmospheric filtering and to causes that are internal to Venus, the planet's overall cratering record differs substantially from that of the other terrestrial planets. Conspicuous in both Venera and Magellan inventories is the deficit of craters with diameters smaller than about 30 kilometers and, in Magellan data, the absence of primary craters with diameters smaller than about 2 kilometers. Small venusian craters tend to differ from craters elsewhere in the solar system. Large craters possess attributes that are shared by craters of comparable sizes on other planets, but they also have qualities that are peculiar to Venus. Furthermore, characteristic of the environments on the surfaces of terrestrial planets is the gradual (on the geological time scale) modification, softening, and, in many instances, destruction and disappearance of topographic detail. Compared to the other planets, venusian craters seem to be young and largely unmodified. The analysis of impact craters — their size–frequency distributions, modes of modification and removal, and relationships to local and regional stratigraphic sequences — forms a cornerstone for efforts aimed at reconstruction of the planetary geological histories. In this respect, Magellan's images of venusian craters have proven to be extraordinarily valuable.

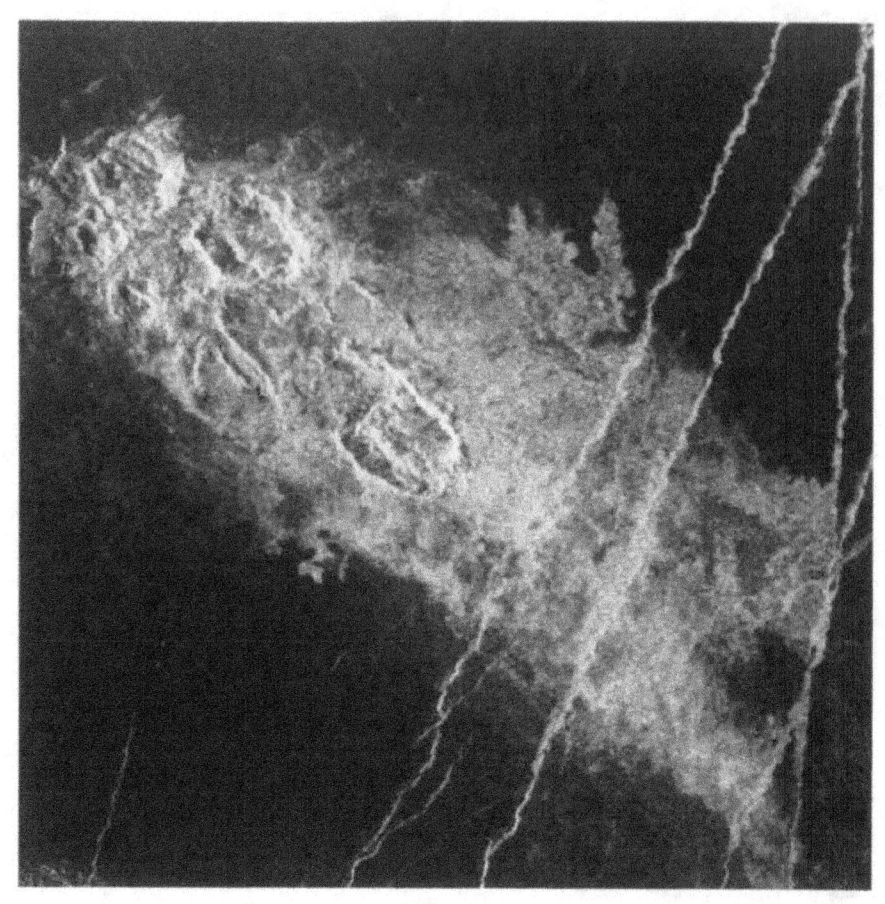

Crater field in Kawelu Planitia (largest diameter 13 km; 45.7°, 253.2°). This unusual crater field deposited ejecta extending to several tens of kilometers. **4.2**

The cratering record of Venus differs substantially from that of the other terrestrial planets

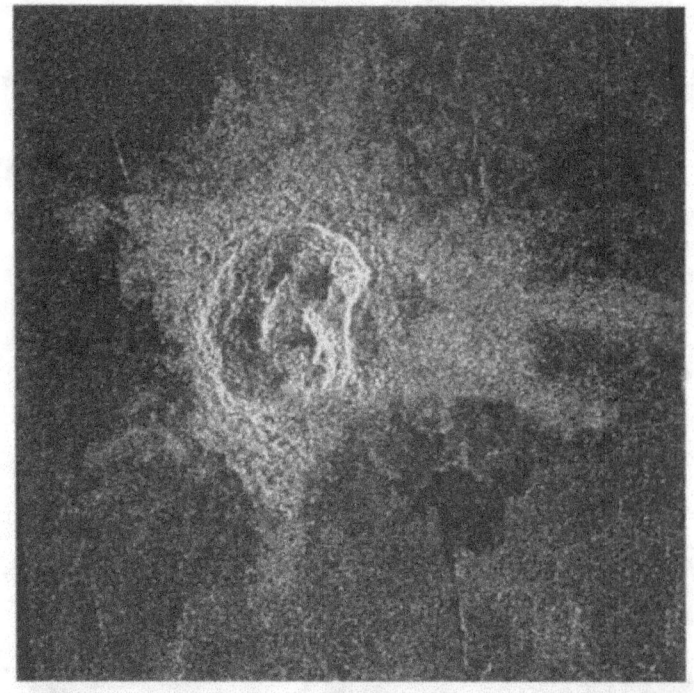

A projectile in the process of disintegration may have created Astrid, an extraordinary, 9 km by 12 km, kidney-shaped crater in the Navka region (−21.4°, 335.2°). The feature is indeed extraordinary because the planforms of impact craters, although not necessarily circular, are convex. 4.4

L ilian, an irregular crater in the Guinevere region (diameter 13.5 km; 25.6°, 336.0°). The crater is an aggregate of four smaller craters, each excavated by a fragment of the common parent bolide. 4.3

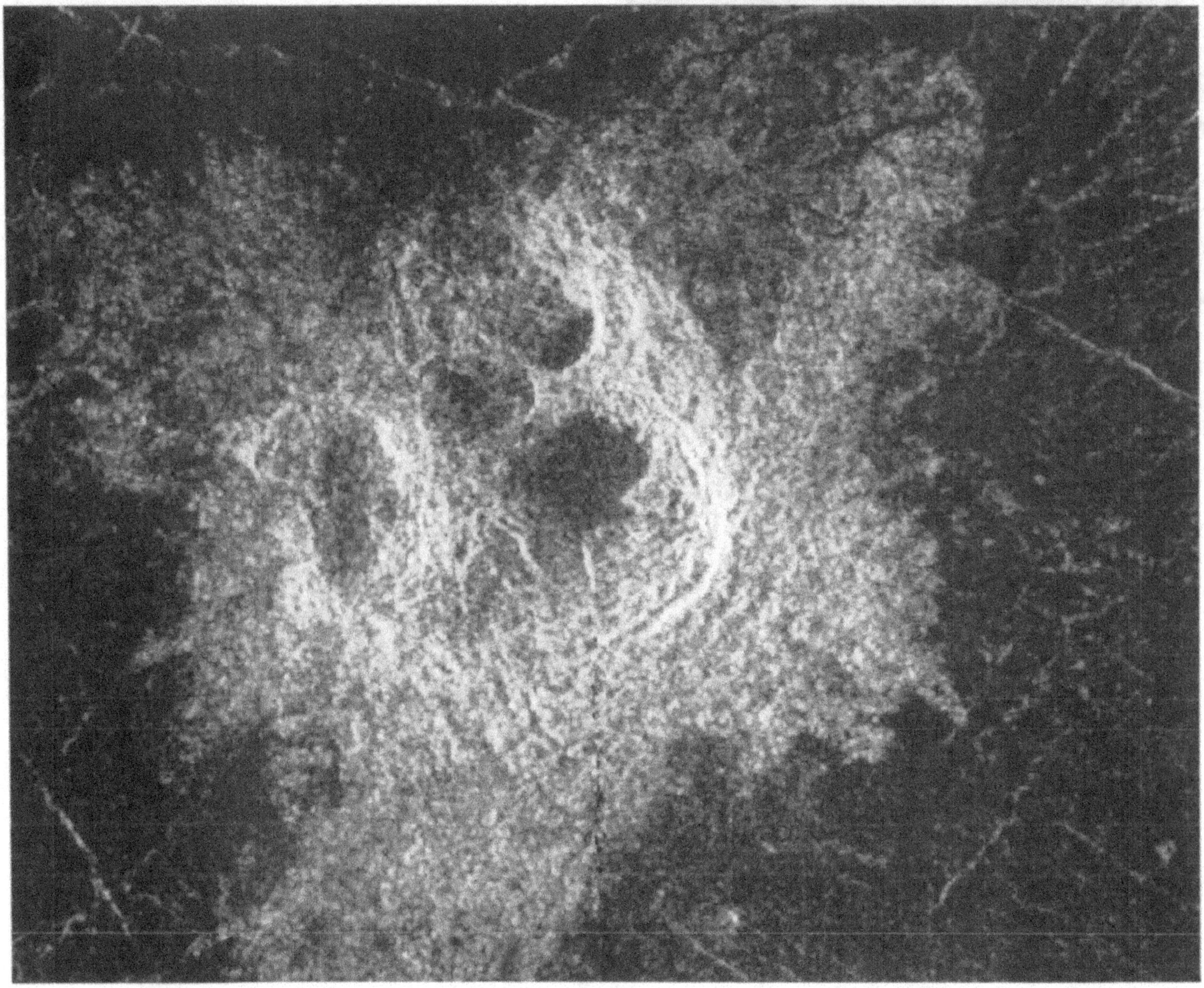

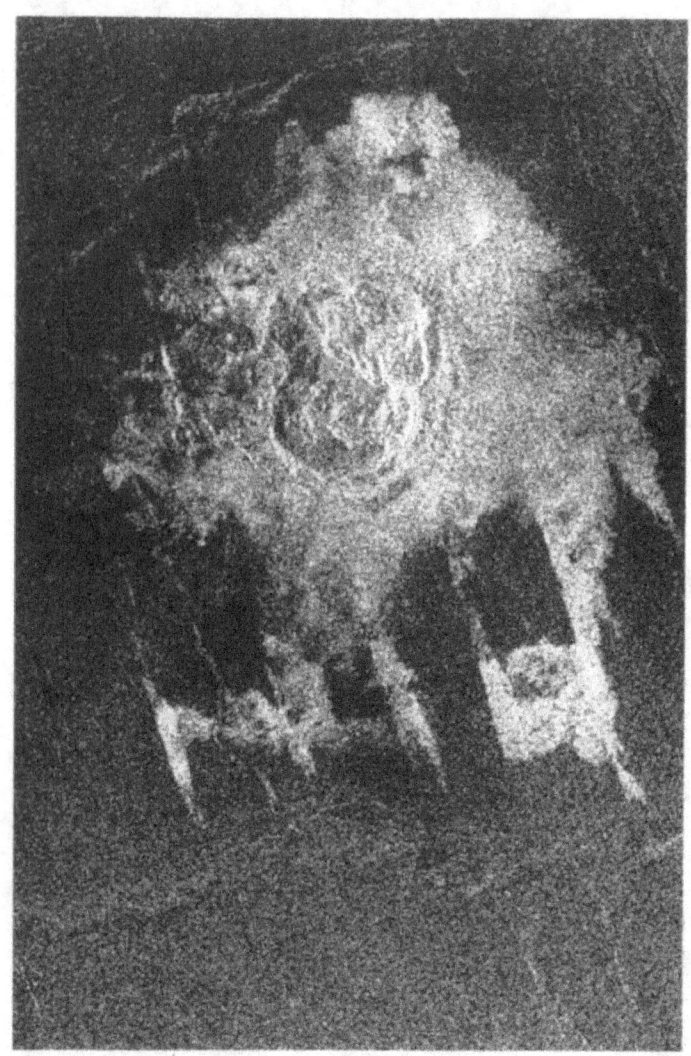

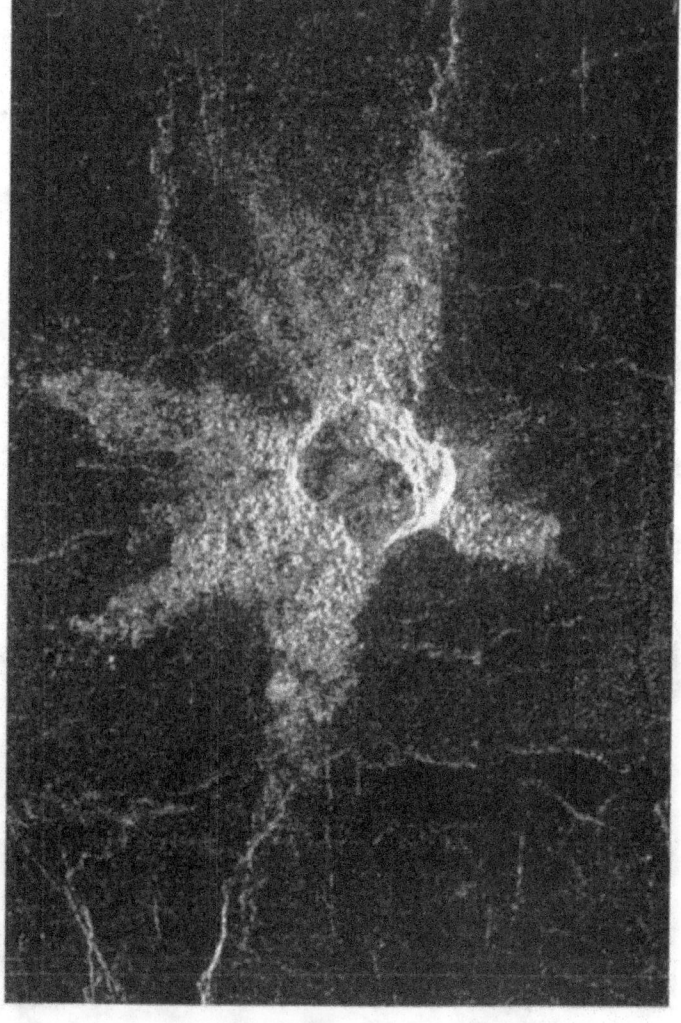

The material ejected from craters formed by impactors traveling along oblique trajectories tends to be distributed in an asymmetric fashion. Such an oblique impact created this unusual, starfish-like ejecta pattern around an irregular crater in Eistla Regio (diameter 6 km; 13.1°, 4.2°). **4.6**

An irregular, multiple-floored crater located south of Sif Mons (diameter 9.2 km; 16.4°, 352.1°). The flows of melted rock generated by the impact of the fragmented meteoroid filled the pre-existing tectonic network. Solidified, the flows formed surfaces more blocky than the underlying places. The resulting roughness contrast made the otherwise invisible boundaries of regional structural blocks detectable by radar. **4.5**

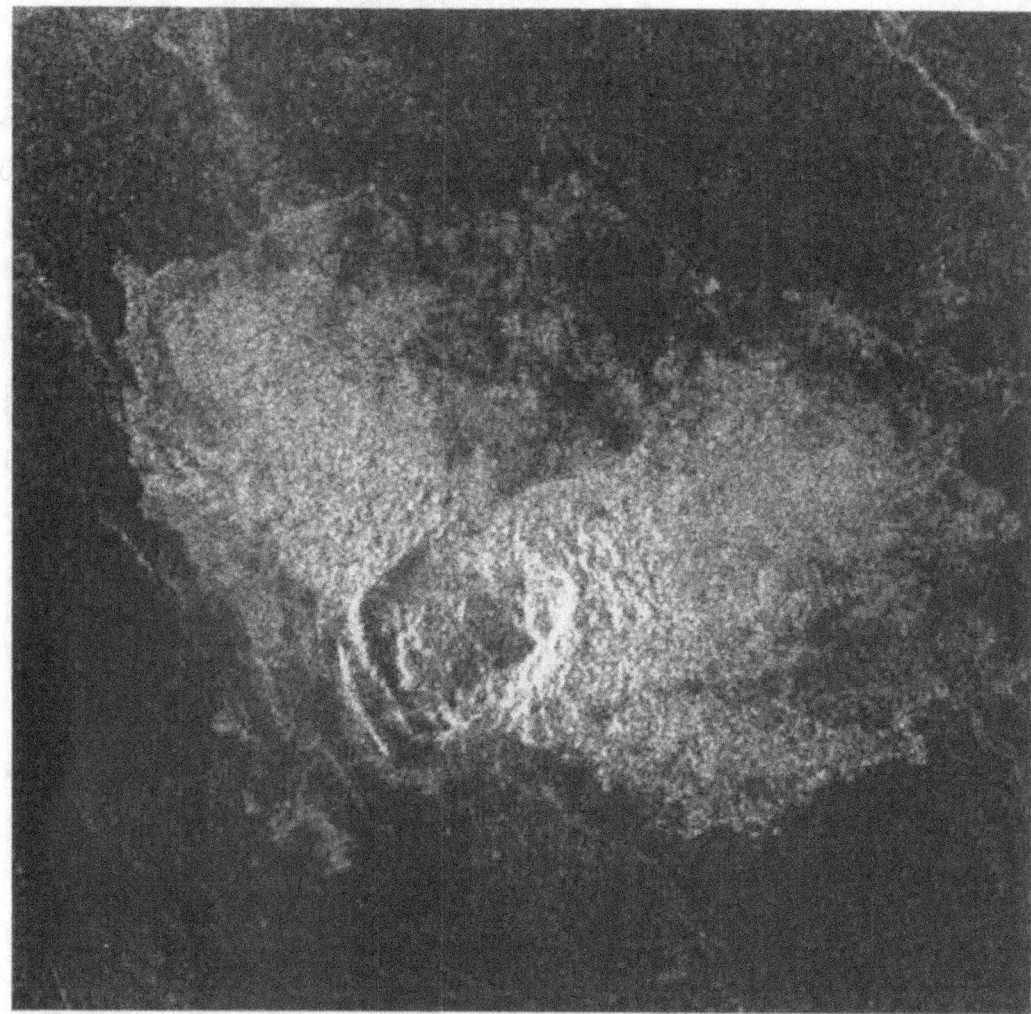

An oblique hit may have generated the butterfly-like ejecta accumulations on both sides of this unnamed crater (diameter 8 km; 11.9°, 352.0°) in the Navka region. 4.7

Curious impact-related patterns, without analogs in the solar system, have been identified in Magellan images. The patterns, involving combinations of dark and bright patches, halos, and streaks, are believed to have been caused by atmospheric shock waves generated by projectiles that did not reach the planet's surface or which disintegrated before reaching it. About 400 cases of such patterns, preferentially at low elevations, have been identified. In a segment of the Lakshmi region (47.0°, 334.0°), featured here, a bright halo surrounds a dark annulus (a "dark margin"), at the center of which is a bright spot. The central bright spot is probably the site where remnants of the disintegrated meteoroid landed. Since the dark radar image signifies a smooth reflecting surface, the shock wave in some manner smoothed the crater neighborhood. It may be that the previously rough surface underwent instant pulverization or that an expansive layer of powdery material was deposited following the impact. Both the ground and the projectile might have been the source of this material. 4.8

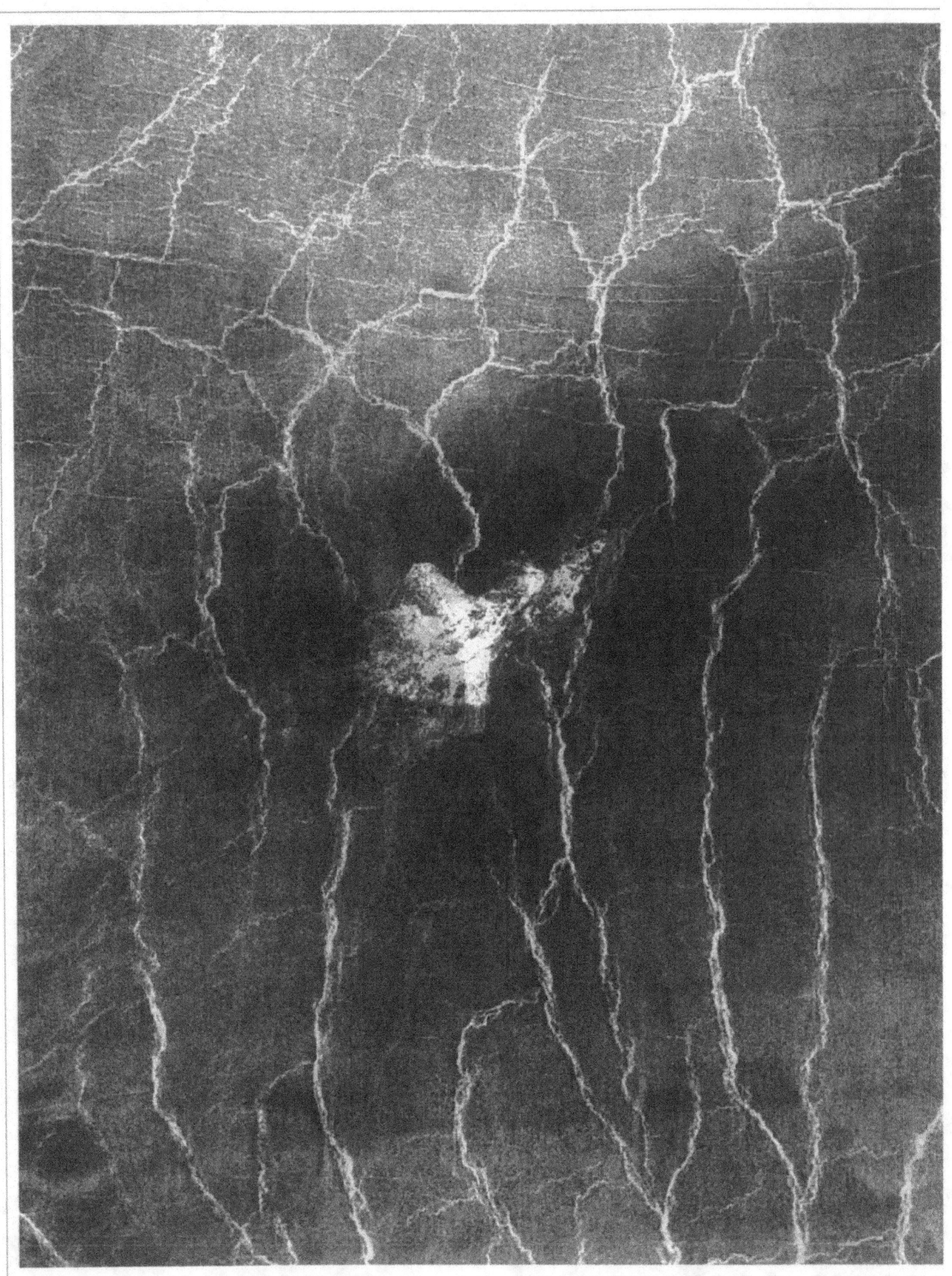

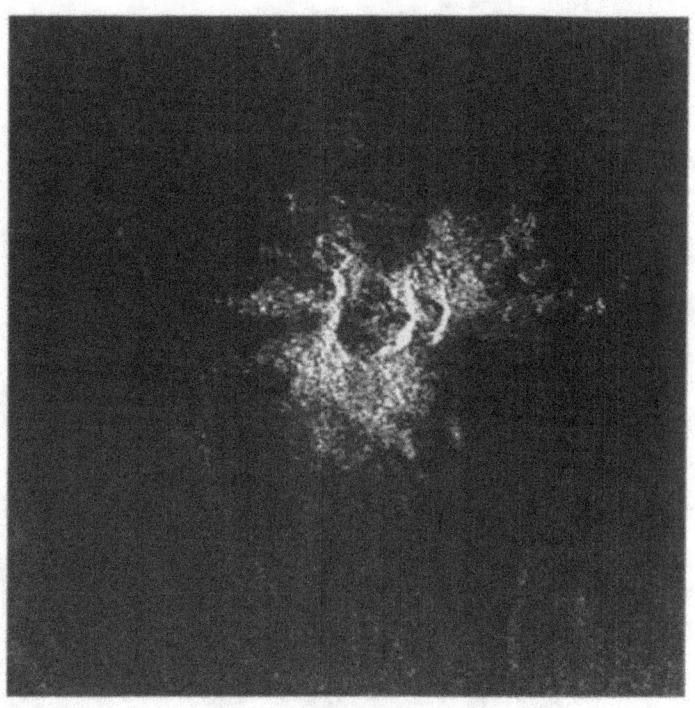

Several noteworthy features are combined in this image (left). A disrupted impactor created a clus-ter of craters (diameter 10 km; 9.3°, 358.0°) in the fractured plains of the Navka region. The impacting swarm struck the surface at an oblique angle, forming butterfly-like ejecta. The resulting atmospheric shock wave generated a dark halo covering a wide area surrounding the impact site. 4.9

Dark halos, which generally have uneven, asymmetric shapes, often extend to distances of up to 10 diameters from the crater rim. In this image, an unnamed multiple crater (diameter 4 km; 55.2°, 350.5°) in Sedna Planitia is encircled by a relatively symmetrical dark halo that stretches out to about five crater diameters. 4.10

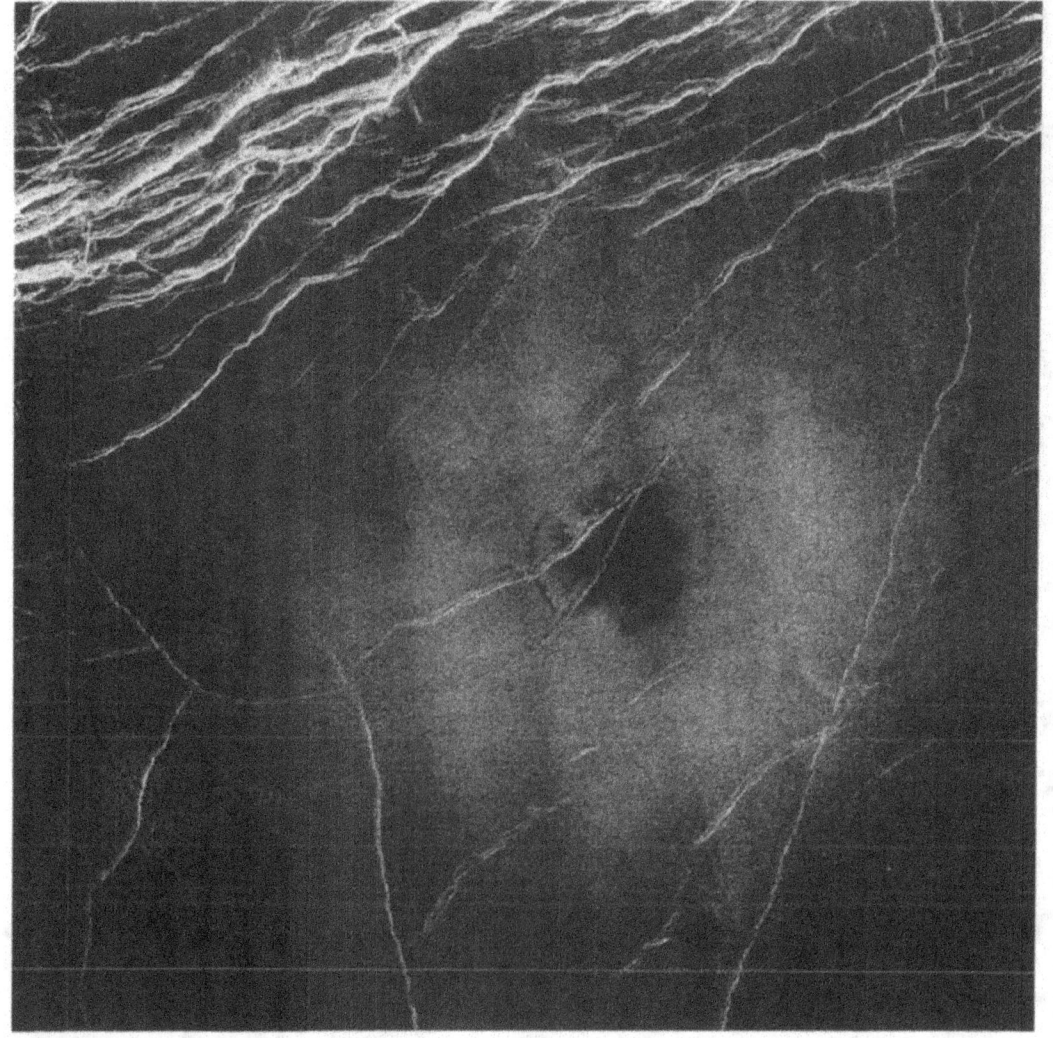

In some cases, the central bright spot inside a dark annulus appears to be absent. The shock wave pre-ceding the impacting meteoroid may have carried sufficient energy to mod-ify the surface and create the dark patch but the projectile itself was de-stroyed while still in transit. The fea-ture in this image has been spotted in Phoebe Regio (0.0°, 270.0°). 4.11

20 km

The dark margin of crater Jeanne (diameter 19.5 km; 40.0°, 331.4°) appears to incorporate smooth materials of two types: lava flows to the southwest and fine-grained sediment to the northeast. The margin has a triangular shape, as does the crater ejecta complex. 4.13

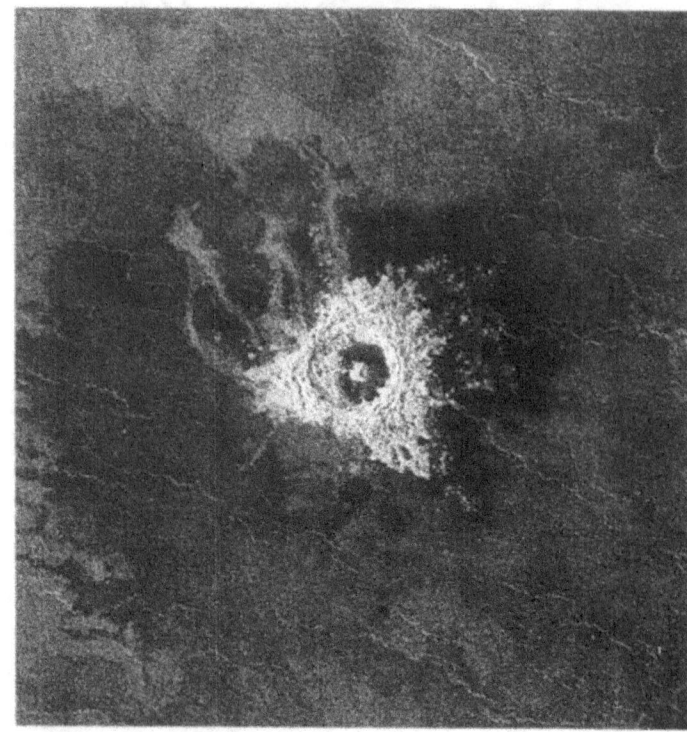

An example of an irregular, dark halo is provided by crater Stephania (diameter 11 km; 51.3°, 333.3°) in Sedna Planitia. The crater was created by an oblique impact. The extent of the halo varies between about two and nine crater diameters. 4.12

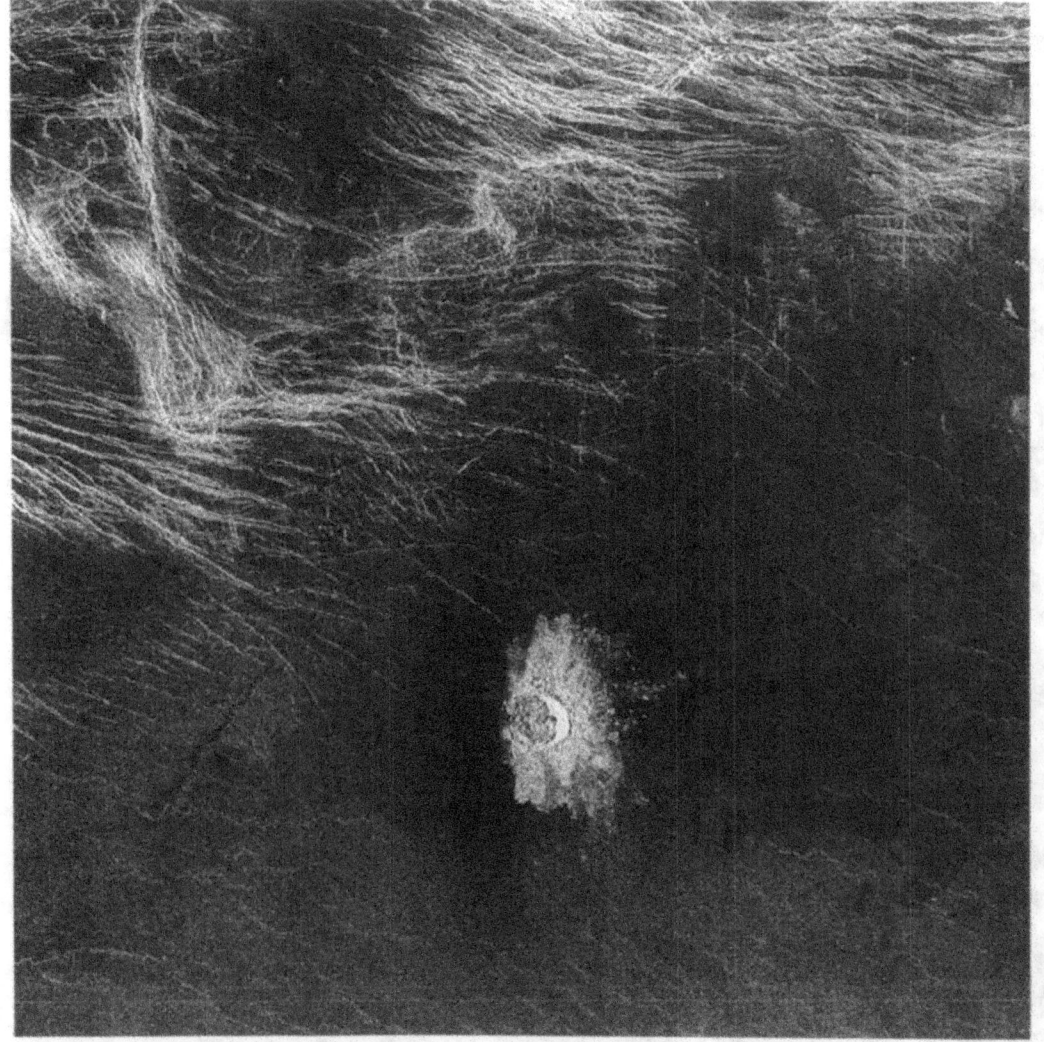

Crater Carson (diameter 41 km; −24.2°, 344.1°), at right, is enveloped by multiple dark halos of an unusual, parabolic shape. Contact between the halos and the ground — unaffected by the crater-producing event — takes place at distances far exceeding 10 crater diameters. The formation and perhaps also the maintenance of open halo arcs may in some way be correlated to the winds near the planet's surface. If winds indeed play a role in the survival of dark halos, then the presence of a halo may indicate the relatively young age of the associated crater. The bright (i.e., strongly reflecting) flows emanating from the crater rim are another distinctively venusian feature. The flows carried material dislodged during the impact to distances far exceeding the reach of the ordinary ballistic emplacement. 4.14

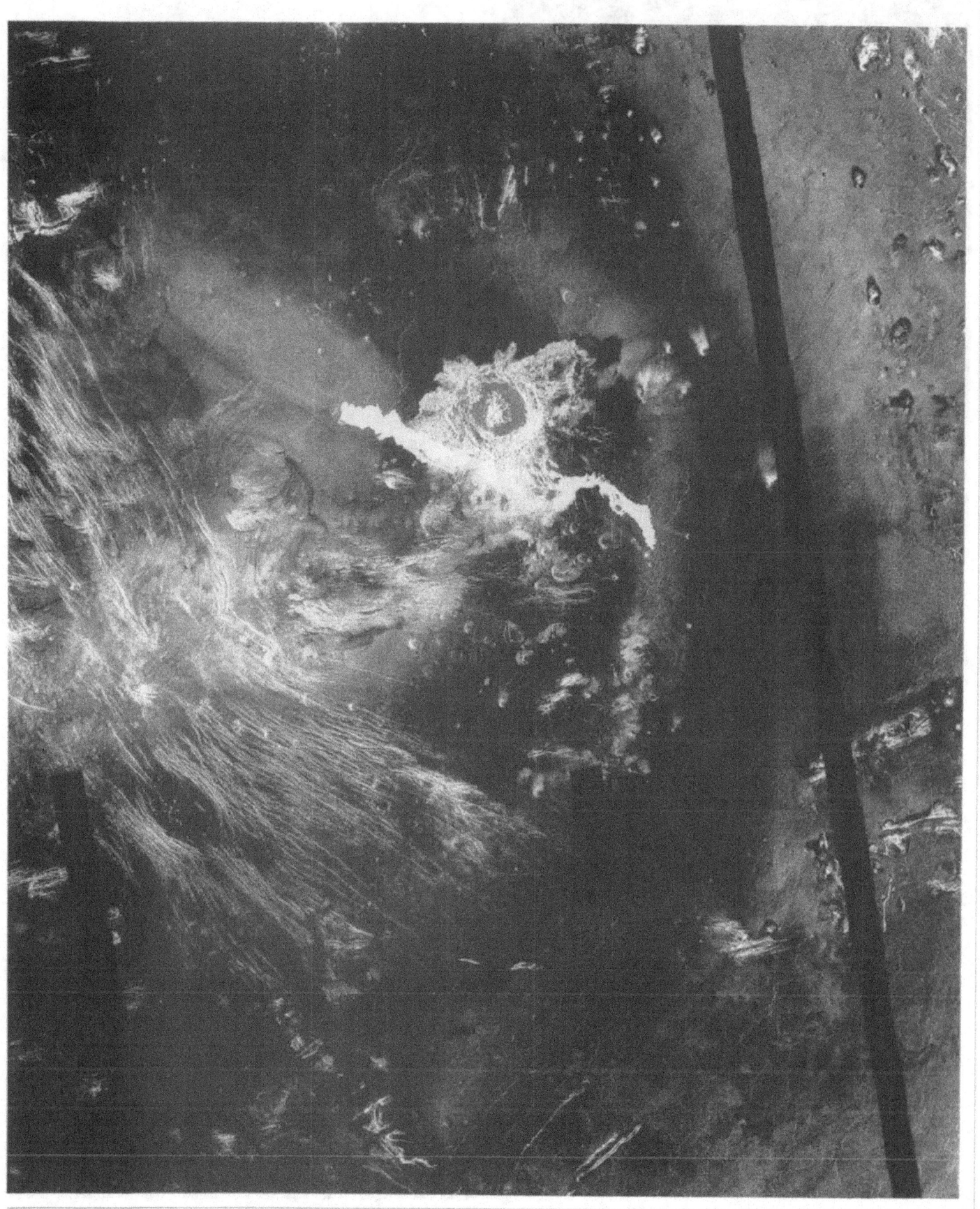

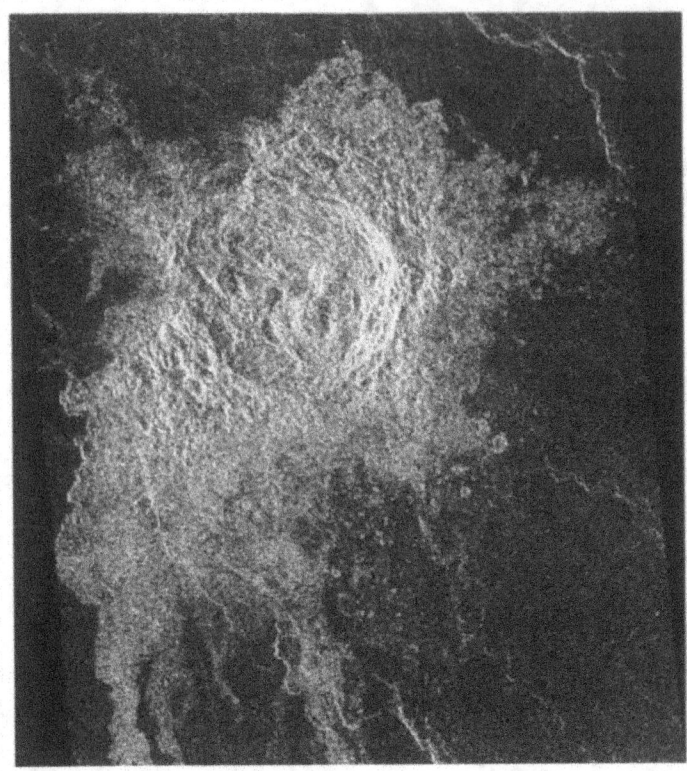

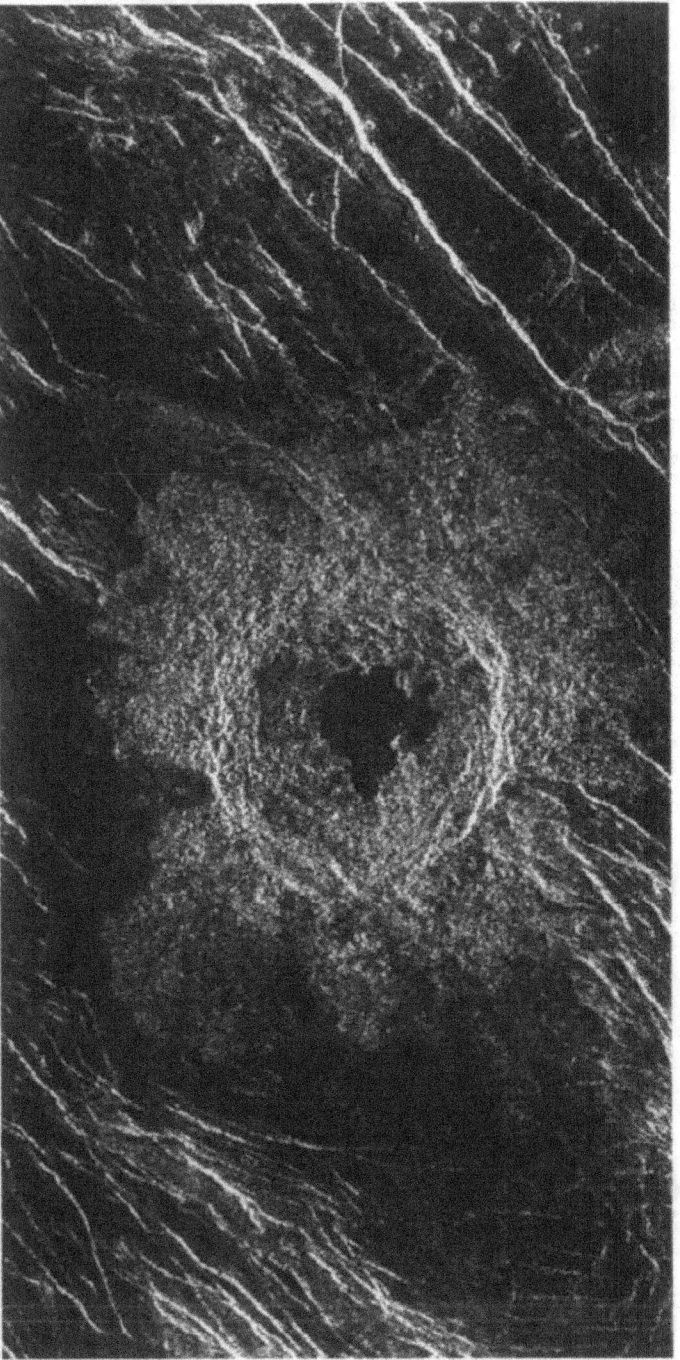

Sabira (diameter 13.5 km; −5.8°, 239.9°) represents a class of craters that have incipient flat floors but still do not exhibit discernible central-peak structures. About 13 percent of craters on Venus share these characteristics. **4.16**

Early in the era of lunar exploration it was noted that craters exhibit a variety of morphologies — from the simple, small bowls created by small impactors, to the enormous, multiring basins that result from collisions with asteroid-size bodies. Craters between the two extreme endpoints of the range of morphologies have a variety of transitional characteristics such as flat floors, single central peaks, multiple central peaks, terraced inner walls, etc. Although Venus lacks simple, bowl-shaped pits, venusian craters do exhibit characteristics associated with complex craters observed on other planets and the Moon. Crater Lydia (diameter 15 km; 10.7°, 340.7°) represents a crater morphology in transition from simple to complex. The crater has not developed a flat floor, but the interior is definitely not bowl-like. Also, the crater outline is largely circular, typical of complex craters. **4.15**

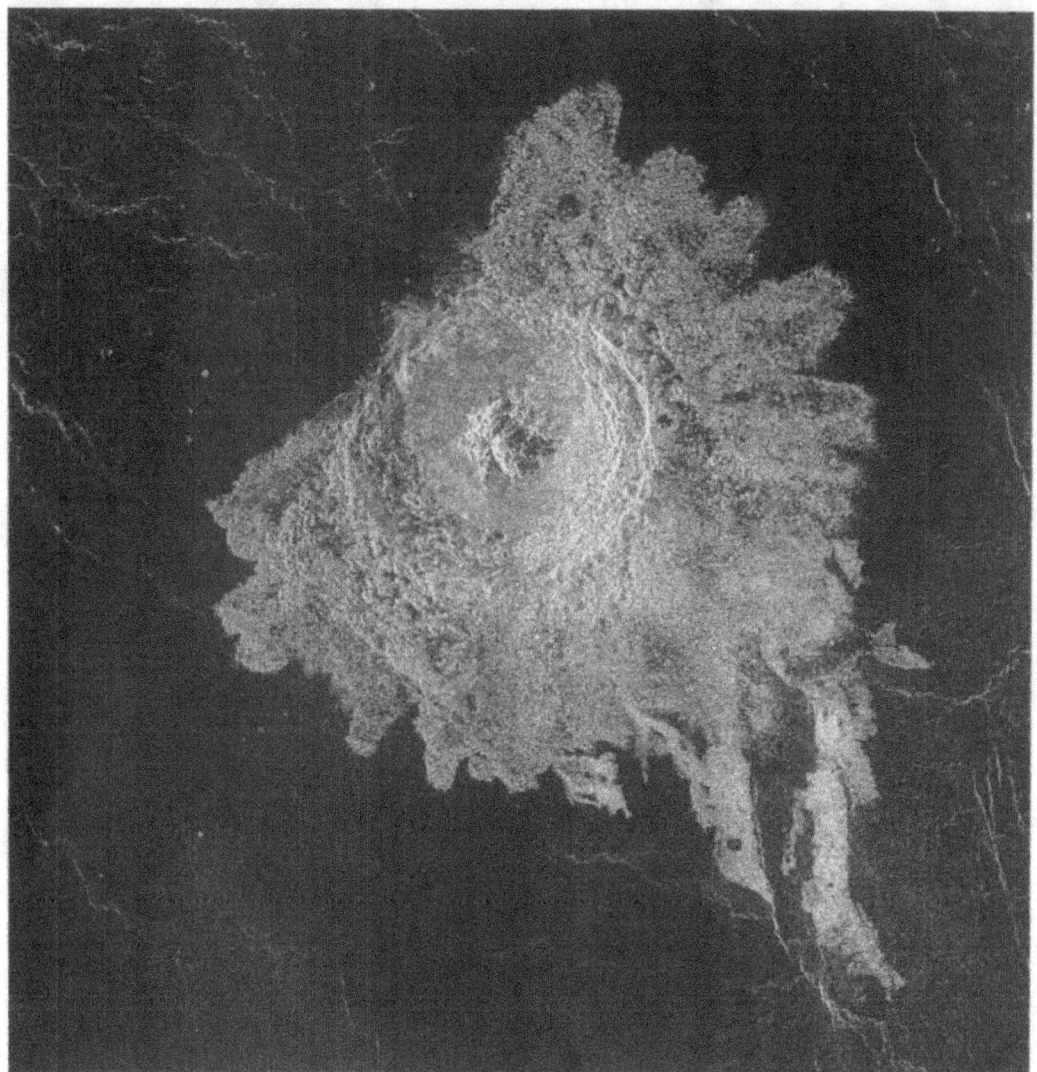

A class of venusian craters with central peaks is exemplified by Aurelia (diameter 31 km; 20.3°, 331.9°). For an impact structure of its size, the crater has an unusually radar-bright floor, suggesting a rugged, blocky texture unmodified by any infilling. About 37 percent of all craters on Venus have central peaks. 4.17

B uck (diameter 21.9 km; −5.8°, 349.6°), a complex crater in Navka Planitia, displays an uncommonly large central peak. 4.18

Venusian craters exhibit characteristics associated with complex craters on other planets and the Moon

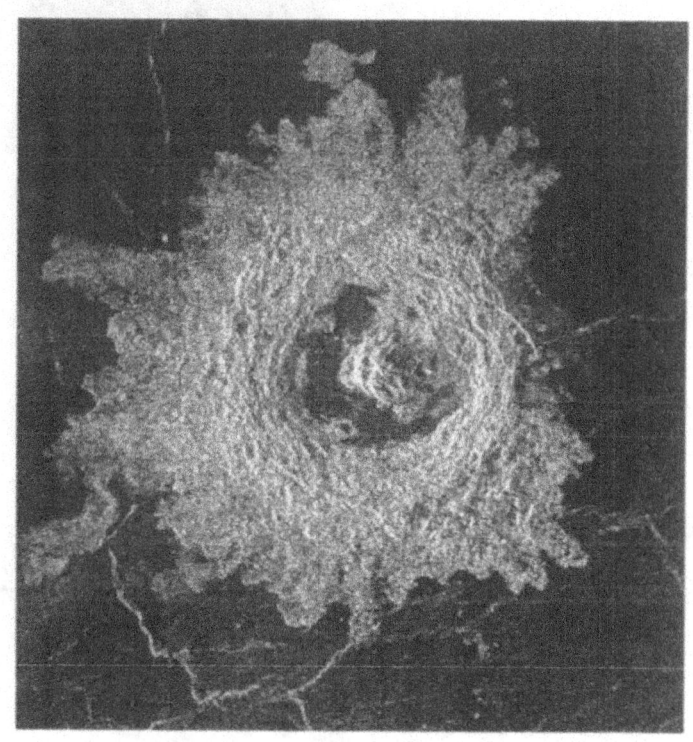

The complex crater Dickinson (diameter 69 km; 74.6°, 177.3°) is a beautiful example of an impact structure in transition from configurations with central peaks to those with central peak rings. The mottled appearance of the crater floor is due to flooding by materials of nonuniform roughness. Absence of ejecta along the rim's western wall indicates a possible impact from the west. 4.19

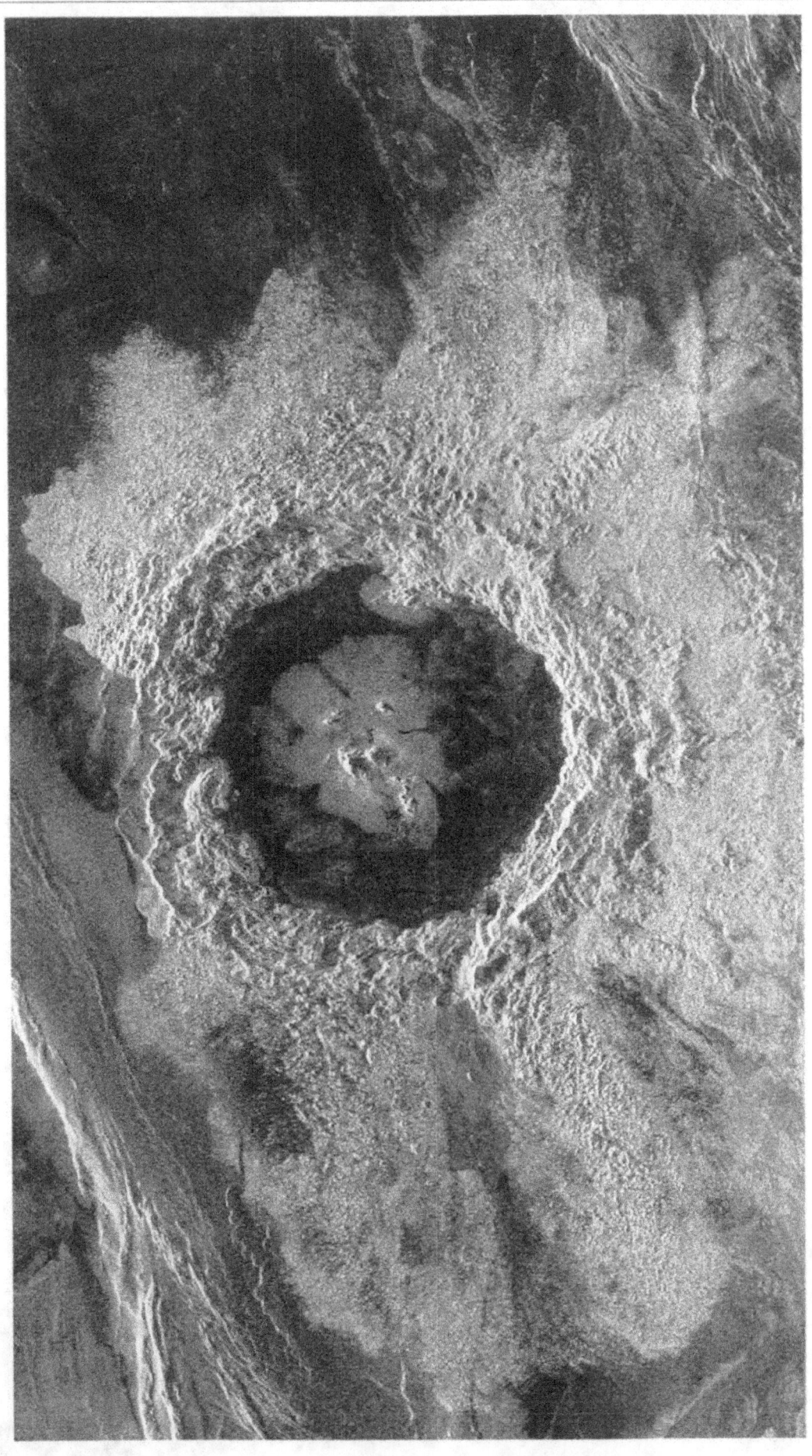

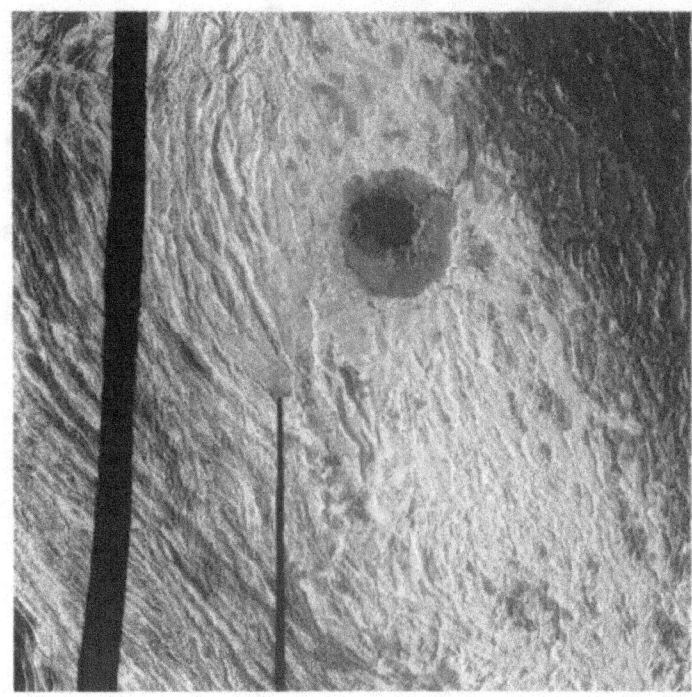

rater Cleopatra (diameter
105 km; 65.9°, 7.0°), first identified in Earth-based radar data,
was initially considered a volcanic
caldera. Venera images challenged this
interpretation, and Magellan images
confirmed Cleopatra's impact origin.
Since the crater appears undeformed
despite its position atop the folded
mountain belt of Maxwell Montes,
the impact may have taken place relatively recently — or the Montes may
have been tectonically inactive for a
considerable period of time. 4.20

case of a crater with an almost
continuous central-peak ring is
represented by Barton (diameter 54 km; 27.4°, 337.5°). The floor
surrounding the ring is radar-smooth,
indicating a probable volcanic infilling. The central-peak rings in venusian craters appear at rim diameters
of about 50 km. 4.21

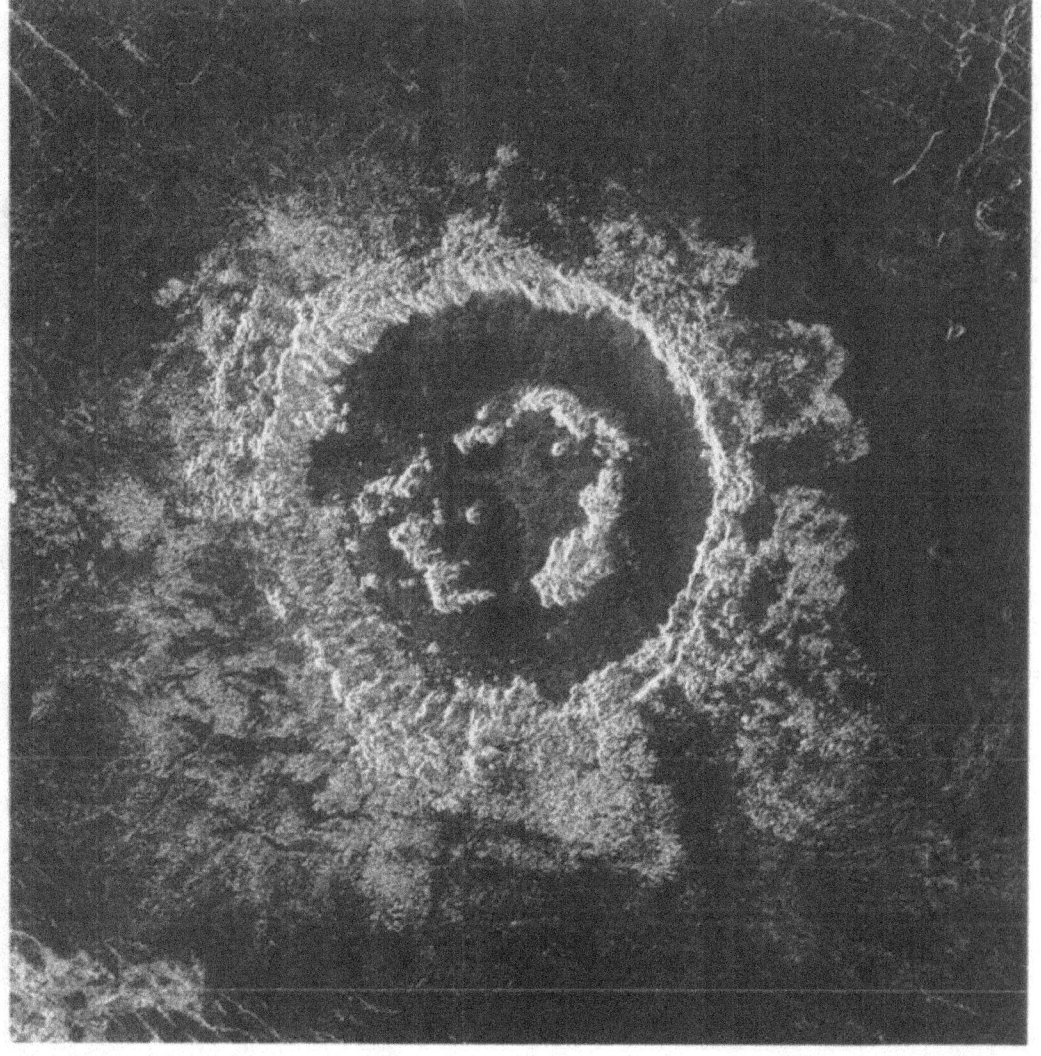

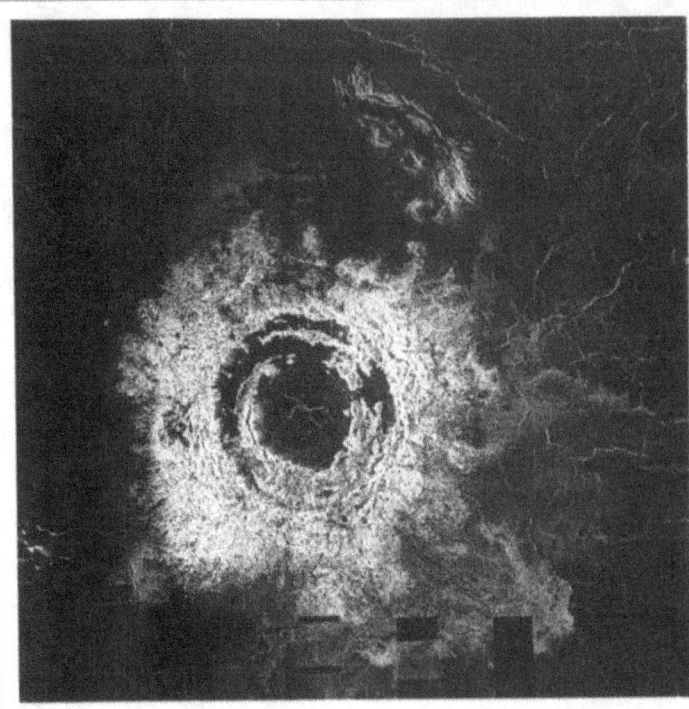

Crater Mead (diameter 280 km; 12.5°, 57.2°), shown at right, north of Aphrodite Terra, is the largest impact crater identified on Venus. The crater is remarkable in having a very well-preserved megaterrace and rim slump terrace — structural elements associated with formation of multiring basins. The boundary of the infilled megaterrace is probably the original rim. [G. G. Schaber, U.S. Geological Survey.] 4.24

Mona Lisa (diameter 86 km; 25.6°, 25.1°), with two topographic rings, is one of the venusian analogs of the multiring basins present on the other terrestrial planets. The crater has a flat, fractured floor and symmetrical ejecta, characteristic of large impact structures. 4.22

Crater Klenova (diameter 140 km; 78.1°, 104.2°) has two well-defined inner rings, but lacks secondary craters. The ejecta outflows, a characteristic usually shared with other large venusian craters, are also missing. Note the small dark-halo crater superposed onto the southeast sector of the ejecta apron. 4.23

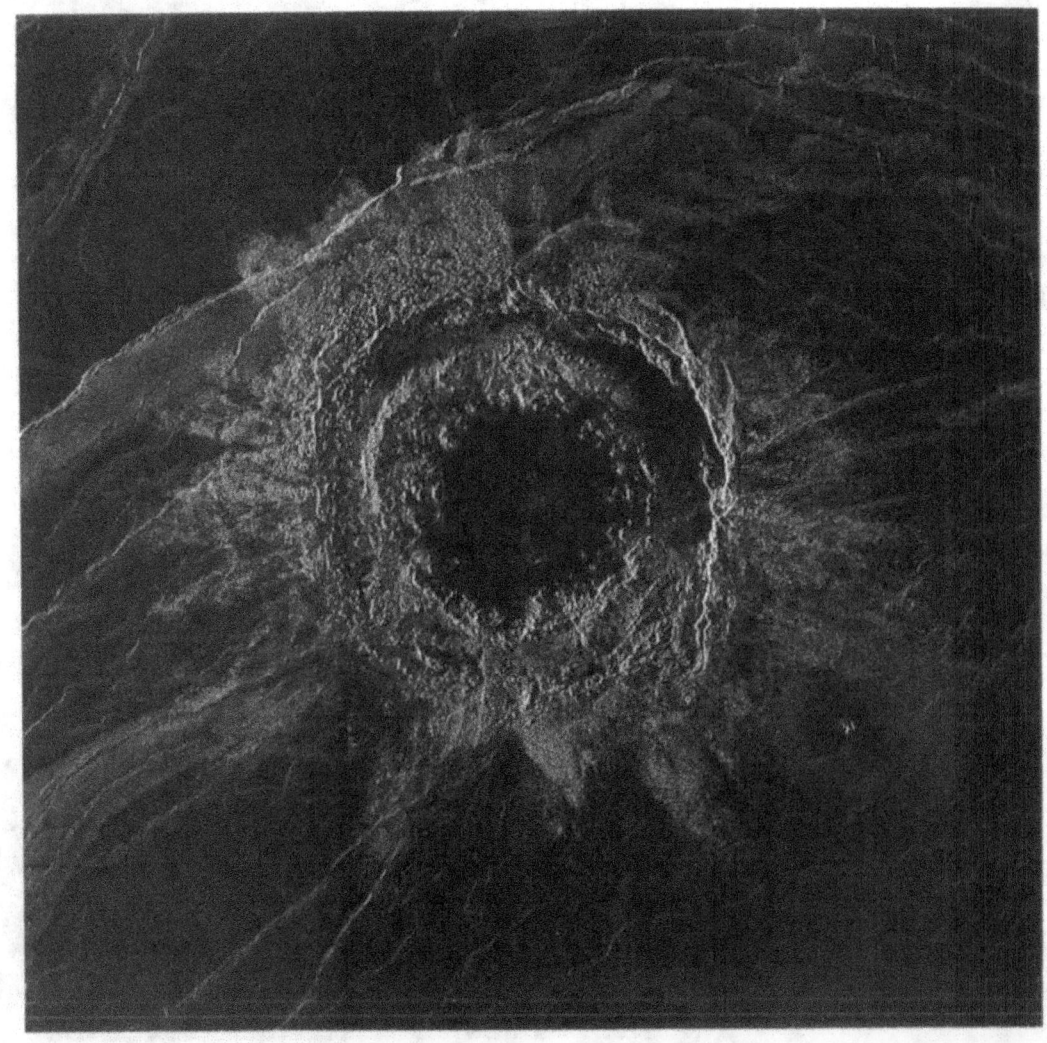

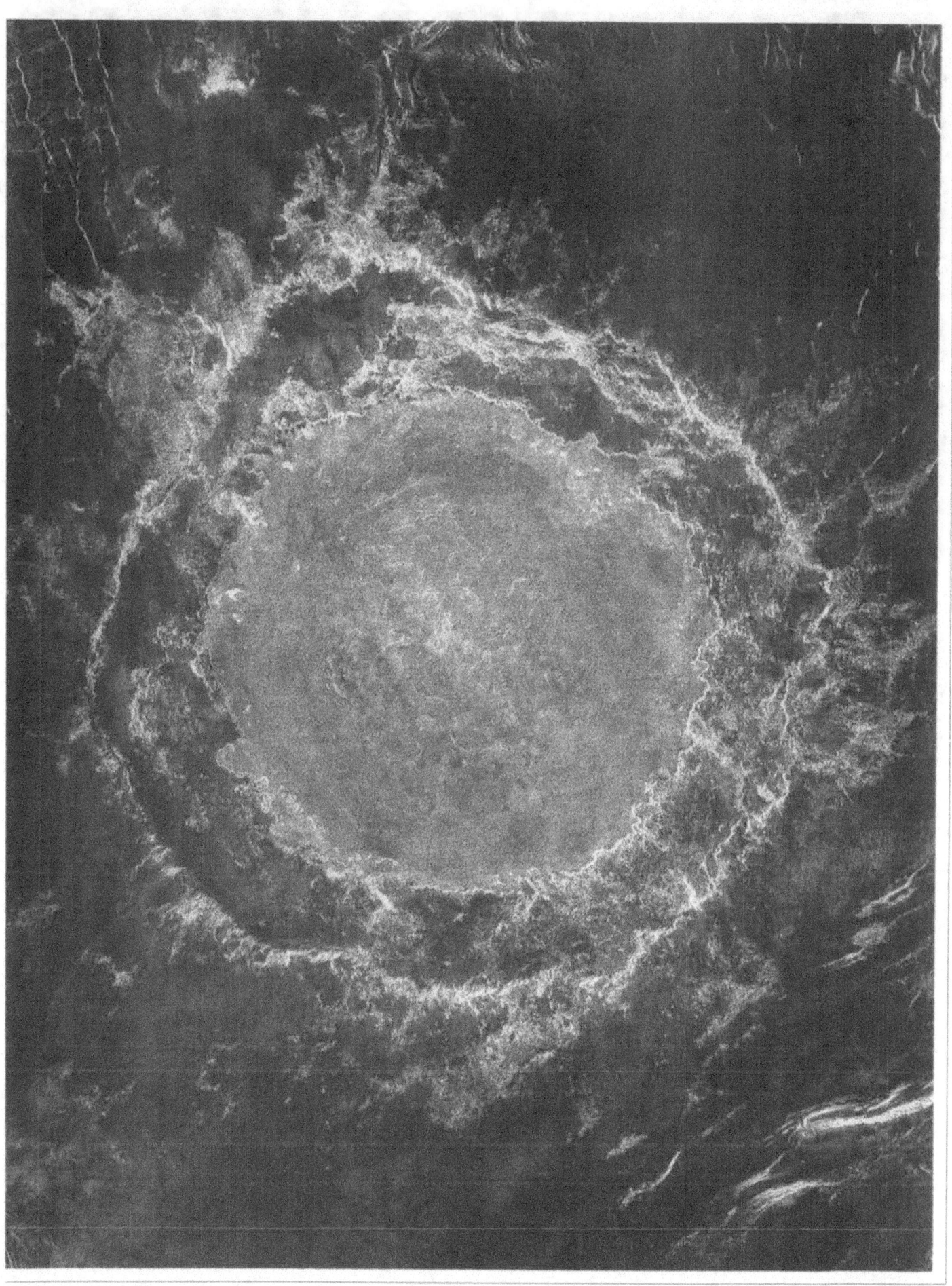

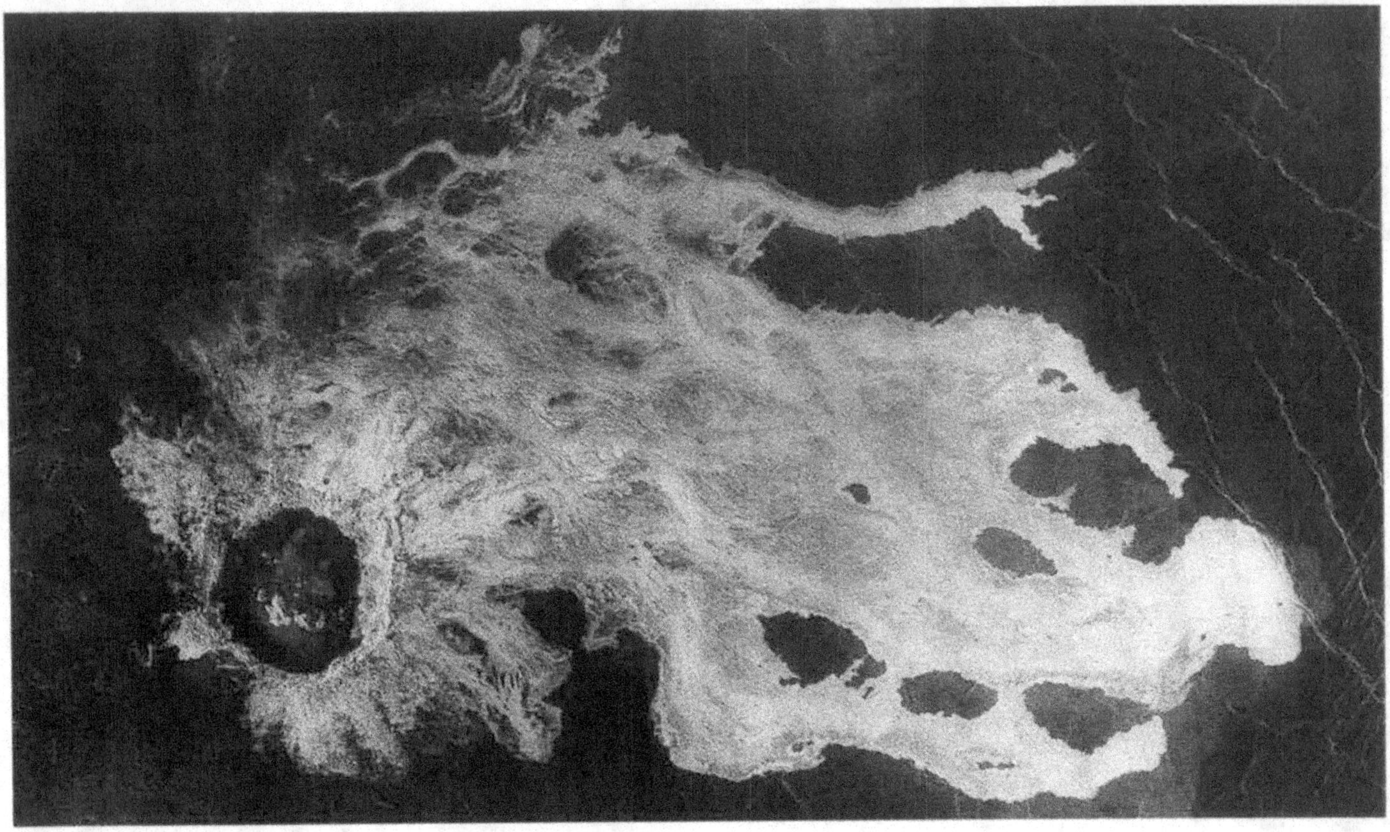

Radar-bright outflows are a distinctive characteristic of many venusian craters. The details of the genesis of these outflows are not clear. High atmospheric pressure and temperature favor the production of melted material at the instant of impact. One possible mode of the emplacement of outflows may involve a gliding motion of packets of melts over an entrapped atmospheric cushion. Since impactors with large kinetic energy produce larger amounts of melt, outflows tend to be more common among large craters. This observation is not a strict rule, however, since the relative occurrence of crater outflows seems to decline for the largest craters. Singular characteristics of crater outflows are their thinness and their length. The outflow deposits of crater Markham (diameter 69.5 km; –4.1°, 155.6°) in Rusalka Planitia extend 370 km from the crater rim. **4.25**

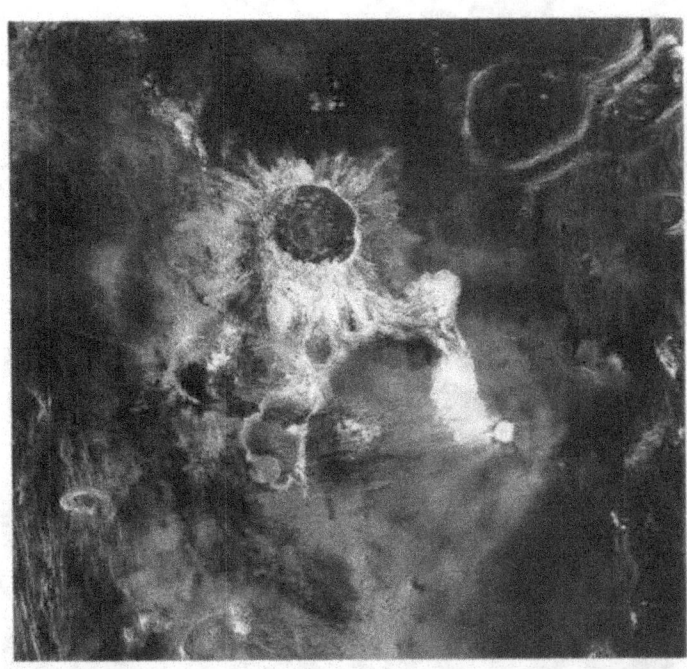

Isabella (diameter 175 km; –29.7°, 204.1°), south of Aphrodite Tetra, is the second largest impact crater on Venus and the largest with outflow deposits. 4.26

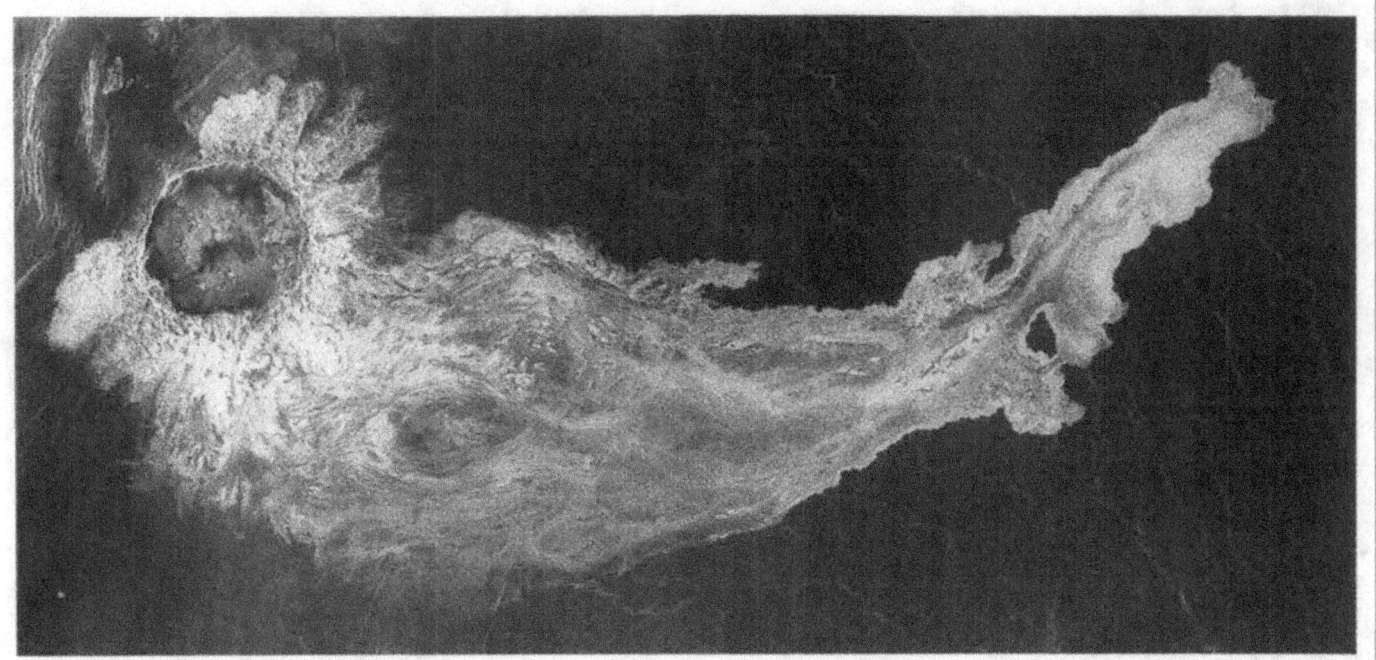

The longest outflows identified on Venus emanate from crater Addams (diameter 90 km; –56.1°, 98.9°) on the eastern edge of Lada Terra. Their farthest reach is 600 km from the crater rim. [*G. G. Schaber, U.S. Geological Survey.*] **4.27**

Riley (diameter 24 km; 14.1°, 72.3°), a medium-size crater in the plains north of Hestia Rupes, exhibits a compact and fairly symmetrical ejecta pattern girded by an extensive, asymmetrical dark margin. **4.28**

The rate of crater-producing collisions depends on the size of
the impacting bodies. Collisions
with larger meteoroids are rare, those
with smaller bodies relatively more
frequent. In rough terms, for each or-
der-of-magnitude decrease in the impactor size, the number of impacts
increases by two orders. An optical
image of a segment of Mare Serenitatis, one of the lunar maria photo-
graphed during the Apollo missions,
demonstrates this phenomenon. The
mare surface is covered by simple crater bowls of a wide range of sizes. The
smallest craters, at the limit of resolution, are the most numerous. Because
the venusian atmosphere does not
permit penetration of small impactors,
surfaces with a cratering record similar to that of Mare Serenitatis do not
exist on Venus. **4.29**

5 km

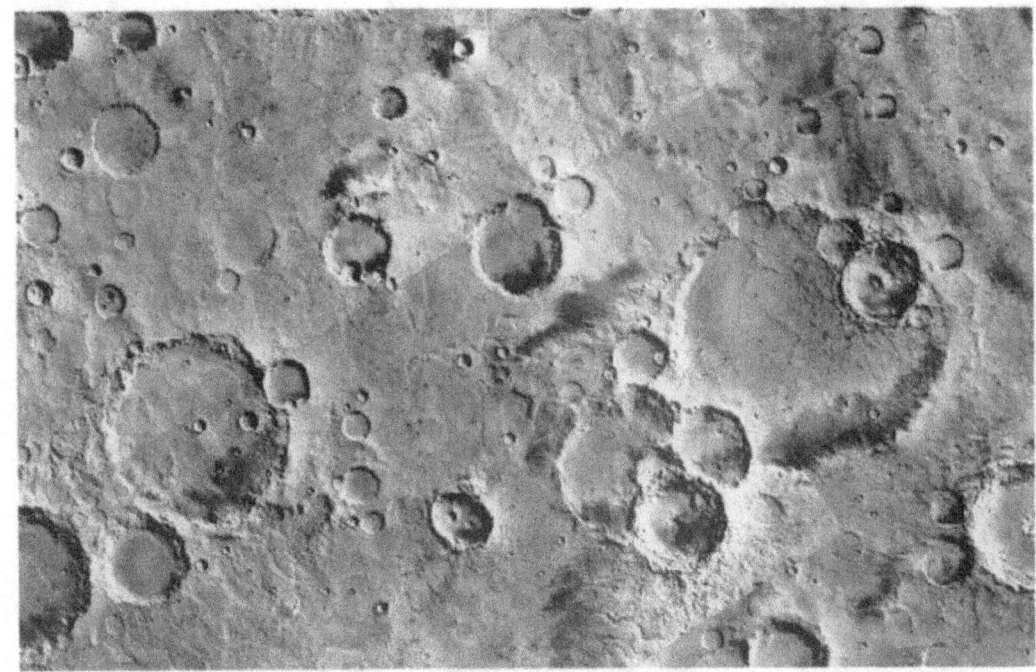

urfaces of terrestrial planets do not stay immutable. A host of processes, both endogenic and exogenic, act through time to change the appearance of the exposed landforms. In simplest terms, weathering sculpts, deposition grades, tectonism, and volcanism alter the planet's very facade. Impact, an exogenic process of marginal significance on Earth, is a major agent of change on bodies with a less aggressive endogenic modification regime. Typical of an impact-modified landscape is the superposition of younger, generally smaller craters on craters that are older and larger. Later impacts (almost) destroy the work of the earlier ones: the new script dominates, but the old one is often discernible. The process is demonstrated by this optical image of the martian cratered highlands. Denning (diameter 140 km; –17.8°, 326.7°), the largest crater in the image, and also the other large craters visible here, date very likely to the era of the late heavy bombardment. The smaller craters, superimposed on Denning, were themselves modified by a succession of subsequent impacts. Massive resurfacing events, probably volcanic in nature, affected the inter-crater plains and obliterated the smaller old craters. The bowl-shaped craters visible in the image are of a relatively recent vintage. 4.30

onditions on the surface of Mercury differ vastly from those prevailing on Mars. The planet has no atmosphere and its crust is thoroughly dessicated. Despite physical differences, the crater modification environment on Mercury bears at least a superficial resemblance to that on Mars. The large craters, created at the end of the era of heavy bombardment, were hammered by the later impacts and suffered the fate similar to that of their martian or lunar counterparts. The scene, which is centered at (–62.0°, 48.0°), is from Mercury's Hermes Trismegisti region 4.31

50 km

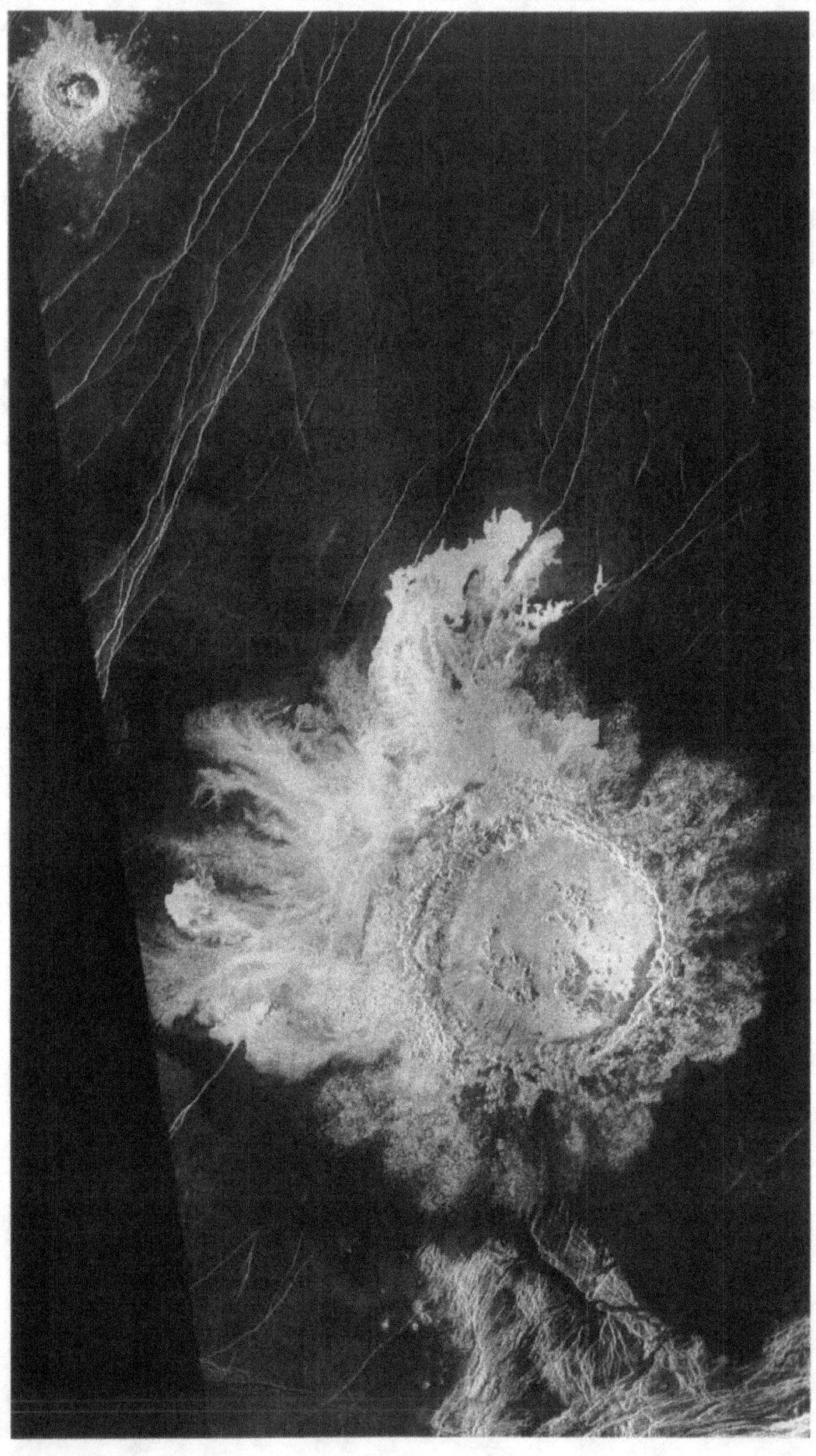

I mages of the cratered landscapes on the Moon, Mars, and Mercury provide a vantage point from which to view Magellan images and appreciate the information they contain. The differences in crater densities between Venus and the other planets and the manner in which craters were altered on the individual planets are striking. Craters on Venus are geographically isolated from each other and have hardly been modified by later impacts. Stuart (diameter 69 km; –30.75°, 20.20°), shown here, can be considered a prototype of a large, well-preserved, venusian impact crater. Unmodified craters, of which this crater is a representative, constitute over 60 percent of all craters on Venus. The remaining approximately 40 percent have been modified exclusively by tectonism and volcanism. 4.32

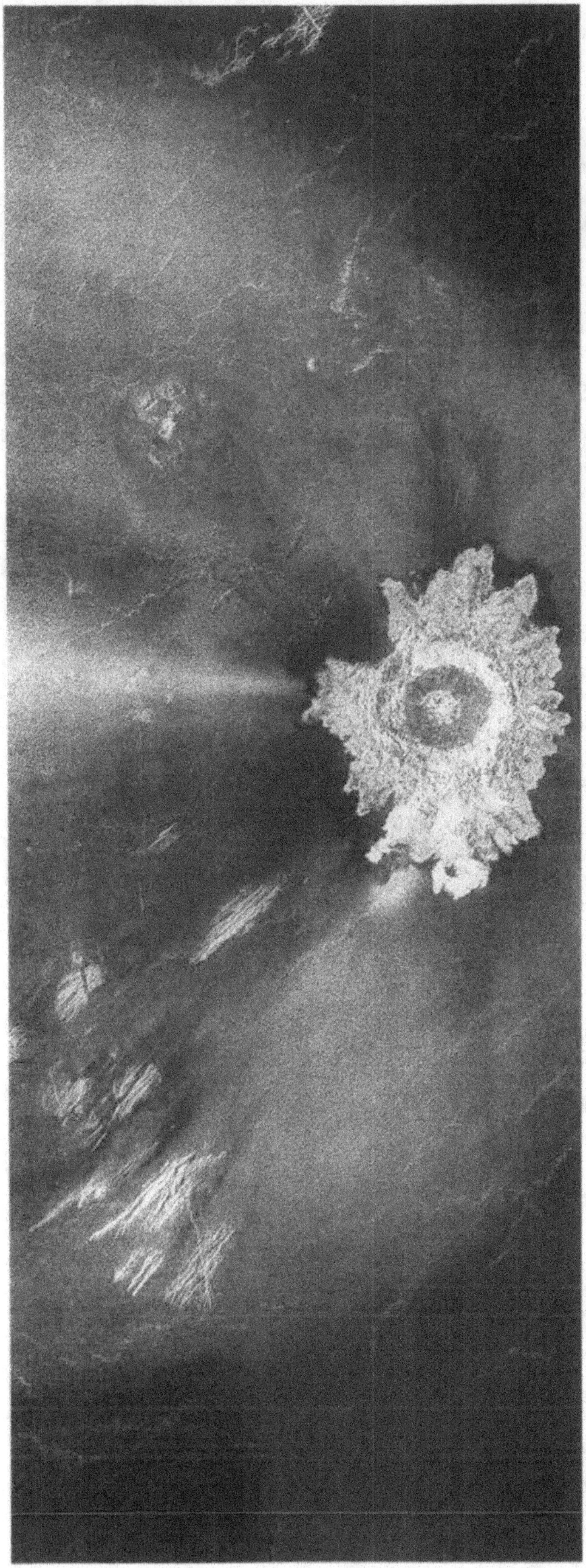

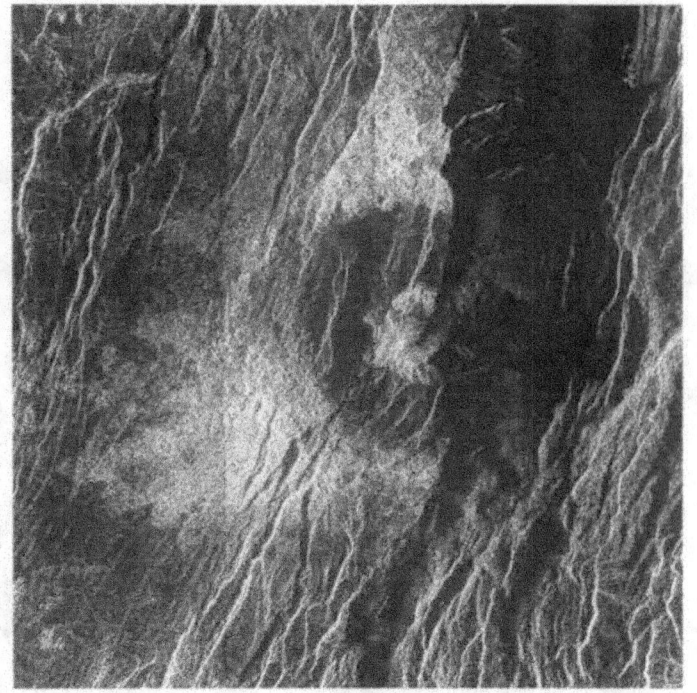

The modified craters on Venus are generally located in areas of tectonic and volcanic activity, such as the Aphrodite volcano tectonic zone. The zone spans the belt from Atla Regio in the east to Ovda Regio in the west. About 35 percent of venusian craters might have been affected by tectonism. Crater Somerville (diameter 37 km; 30.0°, 282.9°), ripped in half in the rift between Rhea and Theia Montes, is perhaps the most dramatic example of the tectonic degradation of venusian craters. One half of the crater is still extant, while the other was demolished by the opening of a 20-km-wide fault. **4.34**

The collision incident resulting in the excavation of crater Adivar (diameter 31 km; 9.0°, 76.1°) may also have led to the creation of an extraordinary, radar-bright streak extending to a distance of almost 20 crater radii from the edge of the crater. Interaction between material dislodged during the impact and the upper-atmosphere winds probably provided the mechanism through which the streak was created. Landscape modifications exemplified by the Adivar event have not been observed on other planets. **4.33**

F ewer than 5 percent of craters presently visible on the venusian surface have been modified by volcanism that originated from external sources. Crater Zamudio (diameter 19 km; 9.7°, 189.2°) on Sapas Mons at the western fringe of Atla Regio, superimposed on a preexisting regional fracture field, has to a large degree been destroyed by a sequence of volcanic flows. Stratigraphically, the most recent flows are also the roughest and are seen here as the bright tongues touching and enveloping the crater rim. 4.35

The surface age of Venus is estimated to be about 200–600 million years

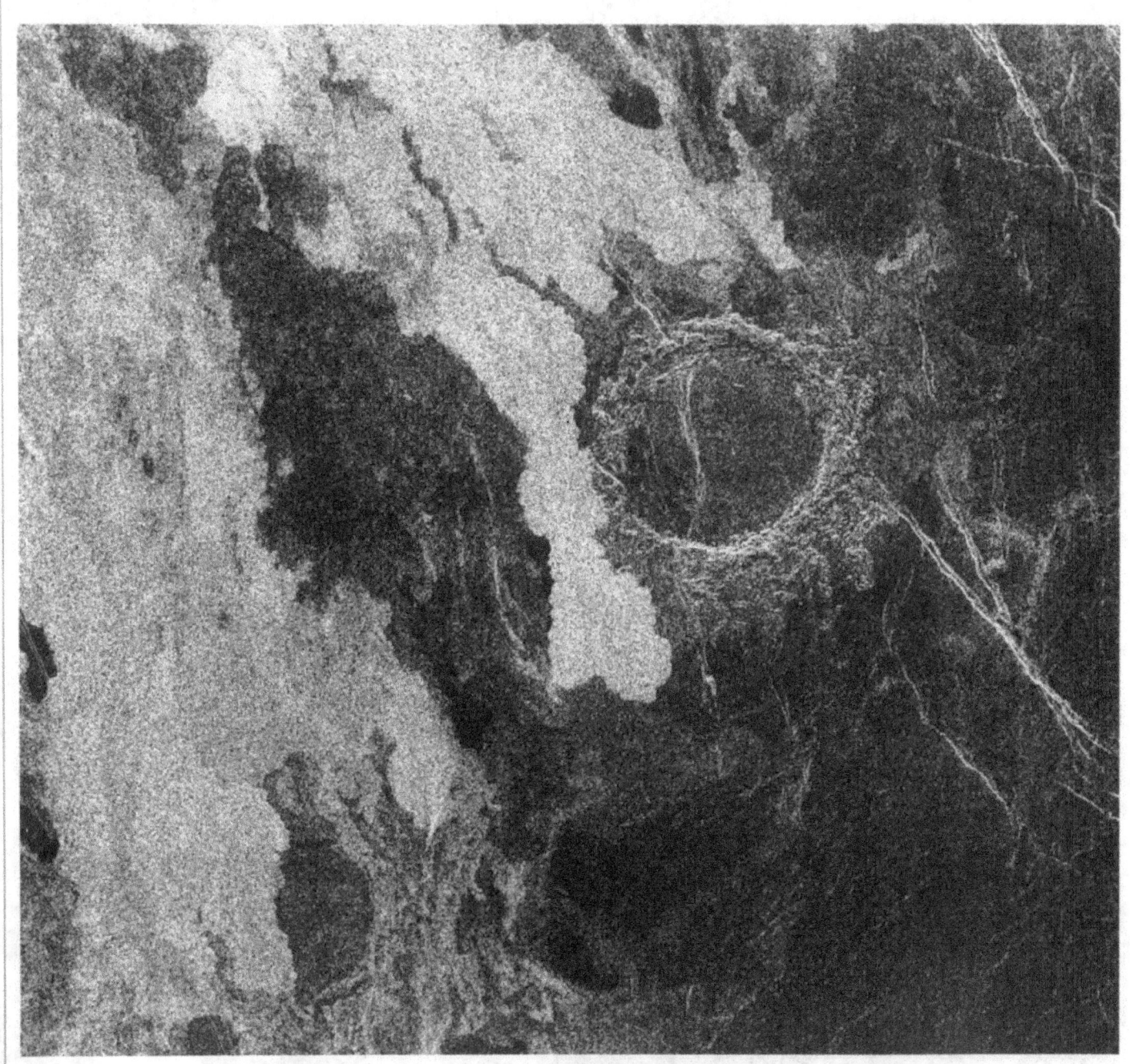

On the average, there is about one impact crater for each 500,000 square km of Venus' surface. (For comparison, the surface area of the state of California amounts to only 400,000 square km.) The scene depicted here covers about 250,000 square km in Lavinia Planitia. It contains three large craters — in venusian terms, an extreme concentration. Clockwise from top left, the craters are Danilova (diameter 49 km; –26.4°, 337.2°), Aglaonice (diameter 65 km; –26.5°, 340.0°), and Saskia (diameter 37 km; –28.6°, 337.1°). Impact craters are more-or-less evenly distributed over the surface of Venus. Their aggregate size–frequency distribution, combined with the estimated meteoroid flux in the planet's vicinity, yields 200–600 million years as the approximate average age of the surface. According to one model, about 300 million years ago, the planet passed through a period of geological activity, in the course of which the previous cratering record was completely erased. Since this cataclysmic surface renewal, accomplished mostly by volcanism, the levels of tectonic and especially volcanic activity declined to an effectively quiescent stage. This scenario implies a sudden expenditure of a large amount of energy by the planet's heat engine, and a certain nonuniformity in the performance of this engine. The competing hypothesis maintains that there are significant regional deviations from the globally uniform crater distribution. Deviations in distributions, if real, would indicate any or all of the following: a wide spread of the surface ages, a non-negligible level of continuing volcanic activity, and a steadier churning of the planet's tectonic machine. Further work may decide in favor of the cataclysmic resurfacing hypothesis; while the debate continues, we can be certain that the craters imaged by Magellan came into being since the last overhaul of the topmost layer of Venus. 4.36

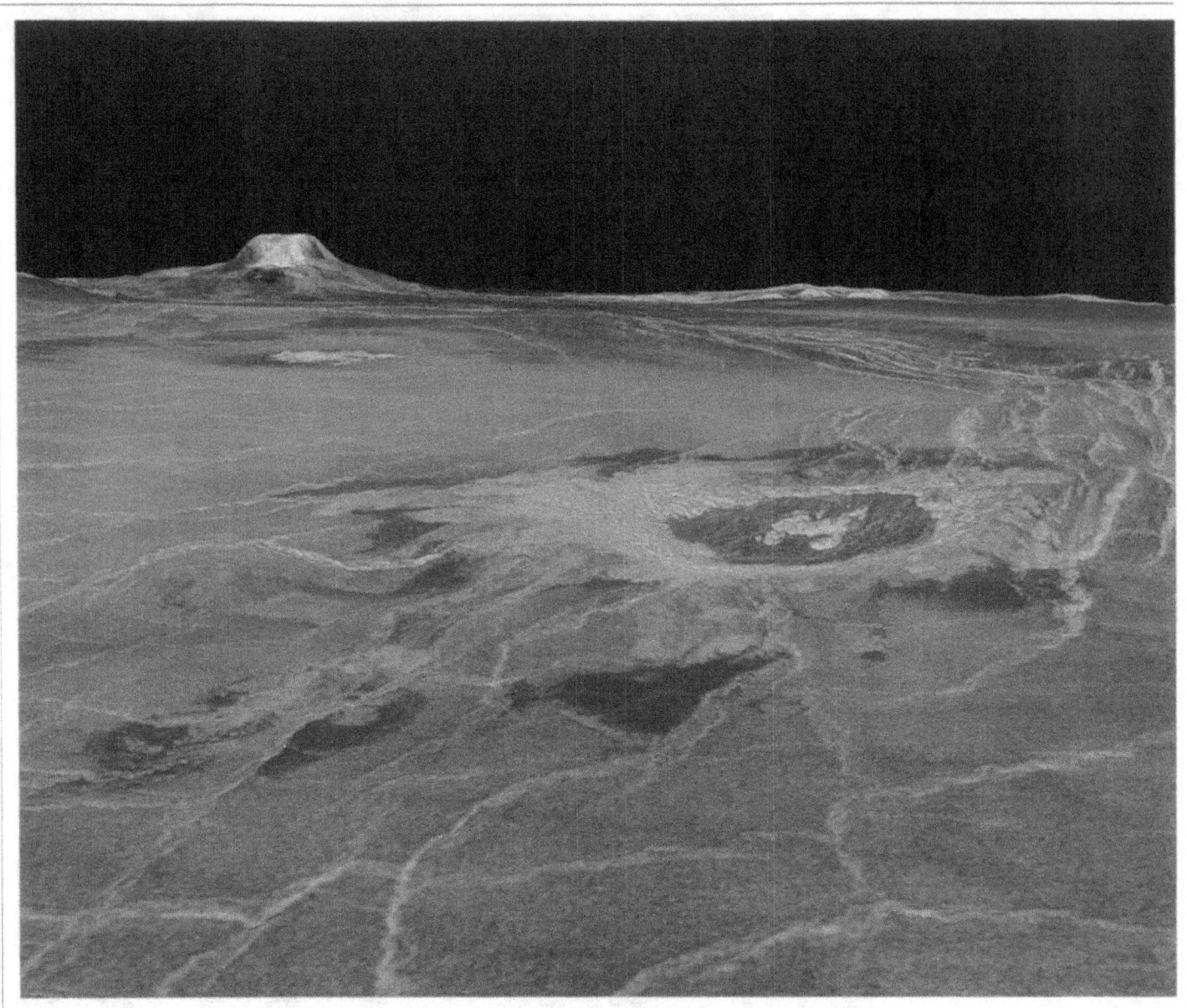

Three-dimensional images, produced by combining Magellan imaging data (including the radar brightness information), Magellan or Pioneer Venus altimetry data, and the color information contained in the optical images taken by Venera 13 and 14 landers, offer arresting views of the venusian landscape. In this vista of the western Eistla Regio, the impact crater Cunitz (diameter 53 km; 14.5°, 350.9°) is seen in the foreground and the volcano Gula Mons (height 3 km; 21.9°, 359.1°) on the horizon. The point from which the scene is viewed is located at an elevation of 0.78 km, about 1300 km southwest of Gula Mons and 215 km from Cunitz. The direction of the view is to the northeast. Vertical exaggeration is 23 times that of the actual terrain. Three-dimensional views of the venusian surface are also featured in Figs. 5.1, 8.1, and 8.15. 4.37

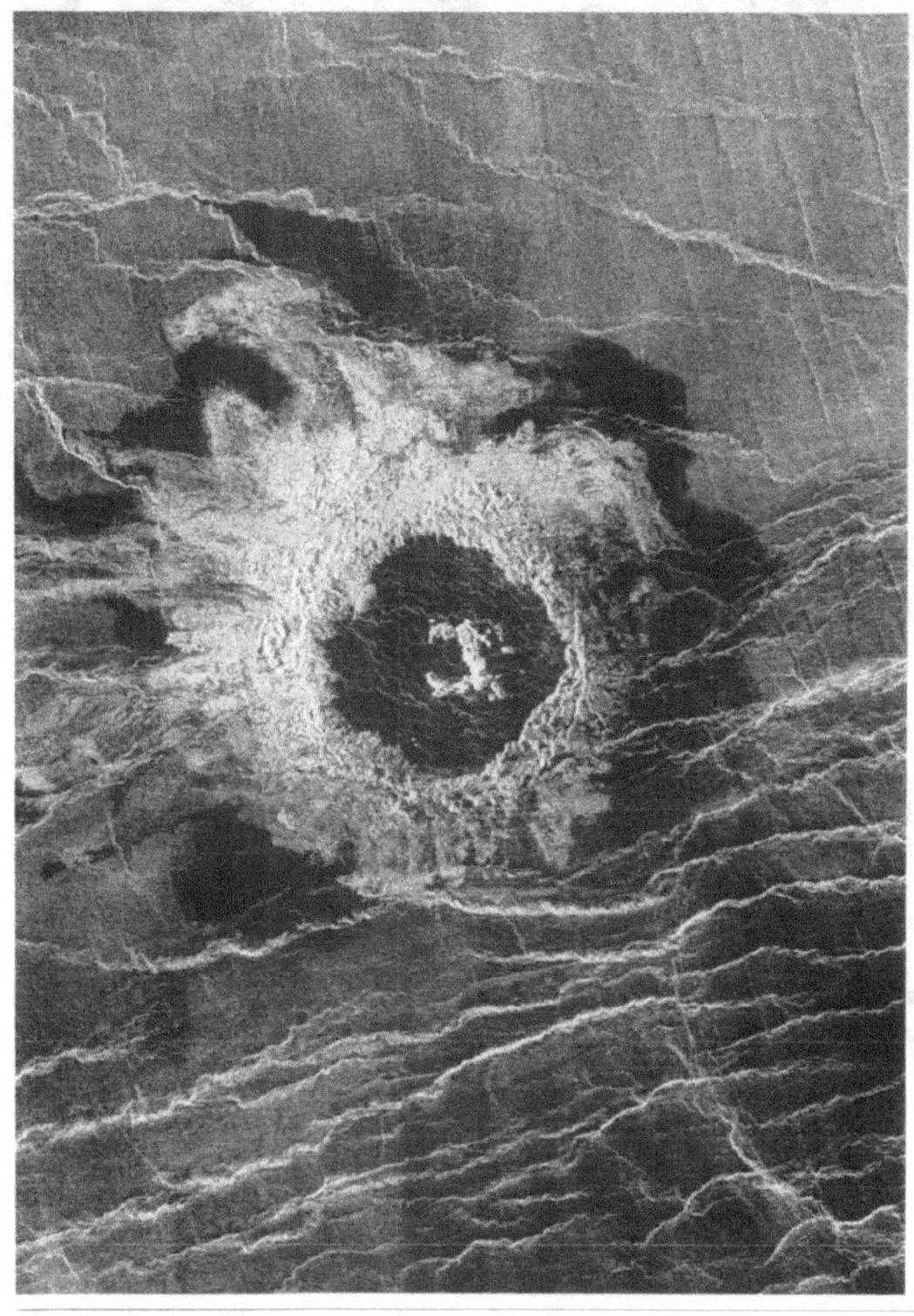

Volcanoes

Vulcan, the Roman deity of fire (from his abode on Mt. Vesuvius), destroyed the towns of Pompeii and Herculaneum in A.D. 79. His Hawaiian sister, Pélé, continues to this day to devour communities bold enough to build in her path. The eruption of volcanoes on Earth's surface has served as the source of myths and faiths, destroyed towns and civilizations, killed thousands upon thousands of Earth's people — and created some of the best farmland in the world.

The large volcano Maat Mons, located in Atla Regio at (1.0°, 194.5°), is shown here in an oblique view, looking to the east. Maat is the tallest shield volcano on Venus, rising 8 km above the planet's mean elevation, 6 km above the local topography, and spanning 500 km across at the base. In the foreground is a large, bright lava flow that extends approximately 600 km from the volcano. Note also the circular impact crater (25 km in diameter) left unmodified by the flow. On the volcano's flanks, more bright lava flows can be seen radiating from near the summit area. In the background on the far left is Ozza Mons, another large shield volcano. This image was created by combining Magellan radar images and topography data. A computer created the projected view; the topographic relief is exaggerated 23 times in order to enhance small features and thus aid in structural analysis. This type of image permits rapid assessment of the relationship between visual features (such as lava flows) and topography.
5.1

The most common geological process on the solid-surface planets (after impact cratering, Chapter 4) is volcanism. The effusion of massive amounts of dark lava into deep depressions caused by impact craters on the Moon results in what we call the face of "The Man in the Moon." Regions of smooth plains between craters on Mercury are interpreted to be the result of eruptions of lava. In 1971, as the clouds from a great dust storm that enveloped the surface of Mars gradually subsided, Mariner 9 imaged the summit of Olympus Mons, the tallest volcano in the solar system. Further Mars imaging by Mariner and the Viking spacecraft revealed several huge volcanoes more than 20 kilometers high as well as many smaller volcanoes along with extensive sheets of lava. Voyager's imaging of the satellites of Jupiter revealed the presence of ice volcanism on Ganymede. In the cases of Jupiter's Io and Neptune's Triton, Voyager imaged the only indications of active volcanism somewhere other than Earth.

The transfer of hot material from the interior of a planet to its exterior is the major mechanism for the loss of internal heat. Internal heat results from several processes: heat from the original accretion of the planet or moon; heat produced by the decay of radioactive elements within the planet; heat that results from the movement of material within the planet; and heat that results from the tidal interaction between adjacent masses (such as Io and Jupiter). The outward motion of this heat (to colder regions) is part of the natural and expected evolution of a planet. The way in which this heat loss is accomplished, however, varies from planet to planet. On small bodies, such as the Moon and Mercury, heat loss by conduction is probably the dominant mechanism. On Io, where there are probably several active volcanoes at any given time, heat loss by advection dominates. On Earth, plate tectonics results in the production of thousands of kilometers of new (and hot) ocean crust at the mid-ocean ridges that cools as it spreads away from the ridges to make the ocean floor. In addition to this heat loss from convection, Earth also loses some of its heat by advection, such as occurs with volcanoes in Hawaii.

Because Venus is similar in size to Earth, and thus might exhibit a similar amount of internal heat, scientists have been curious to see how our sister planet has gone about shedding its heat. Analysis of images from the Magellan mission and geophysical modeling have allowed scientists to begin addressing this question. Thus far Venus does not appear to display the plate tectonics operating on Earth. There is, however, a great deal of evidence suggesting extensive volcanism on the surface of Venus. In fact, approximately 80 percent of the surface landforms on Venus can be attributed to some form of volcanic process. The major subdivision in these processes is related to the form of the volcanic deposits. In many cases, volcanism is focused at a distinct source and

the deposits are arranged (often radially) about this source. This type of volcanism (which is called "centralized volcanism" and is the subject of this chapter) results in volcanoes (such as those on Earth) as well as several other, sometimes bizarre, landforms. The second type of volcanism is not as radial or centralized and generally covers much wider regions of the planet with lava flows. These "flood type" eruptions are described in Chapter 6. Both cases present a sampling of the types and various styles of volcanism that have played such an important role in shaping and reshaping the surface of Venus.

Sapas Mons is another large volcano located on the northwestern edge of Atla Regio (8.0°, 188.0°). It is approximately 600 km in diameter and reaches 1.5 km above the surrounding plains. Many radar-bright and radar-dark lava flows extend radially away from the unique twin peaks of this volcano. The slightly high topography associated with the rim of the 40-km impact crater on the northeast flank has deflected the bright lava flows, suggesting that the flows are quite thin. The bright lines radiating away from the volcano (best seen on the southern flanks) are likely fractures related to magma movement beneath the surface. (See Fig. 5.25 for comparison). **5.2**

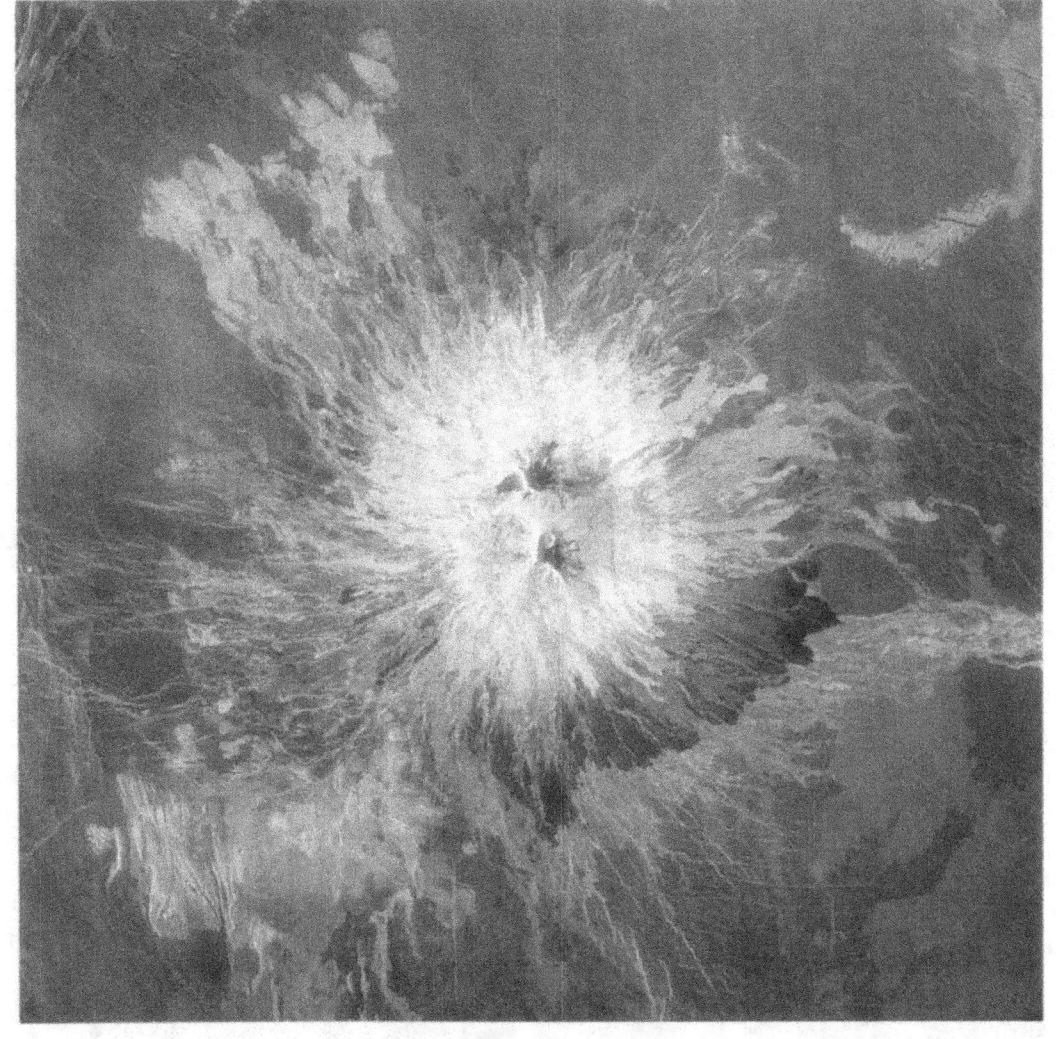

An unnamed volcano (10.0°, 275.0°), located in the plains some 1000 km west of Devana Chasma, rises 1.5 km above the surrounding terrain and is 520 km wide at the base. It differs from many other large volcanoes on Venus in that there are few associated structures, such as fractures, faults, or small cones. Bright lava flows can be seen ponding at the base on the north and west sides of the edifice, suggesting a slight trough or shallowing of topography in the region. **5.3**

The form of volcanism most familiar to Earth-dwellers is the cone-shaped volcano. On Earth, volcanoes are of two general types: relatively quietly erupting shield volcanoes (e.g., those on Hawaii) and explosively erupting composite volcanoes (e.g., Mt. St. Helens and Mt. Pinatubo). The major difference between these two types of volcanoes is the source of the erupting material. Hawaiian-type magmas come from deep within Earth (from areas called "hot spots"), are relatively fluid, and thus easily allow gases to escape. Composite volcanoes are associated with plate tectonics: the release of seawater from the down-going plate at a subduction zone enhances melting and produces sticky magmas that cannot release gases gently and thus often explode. The morphology (long, thin lava flows) and the apparent absence on Venus of either plate tectonics or significant amounts of water suggests that the large volcanoes on Venus are probably more similar to the Hawaiian model. Unlike Hawaiian volcanoes, however, which may grow to only tens of kilometers in width but up to 8 kilometers in height, the large volcanoes on Venus are generally much wider (several hundreds of kilometers) and somewhat shorter (an average height of only about 1.5 kilometers).

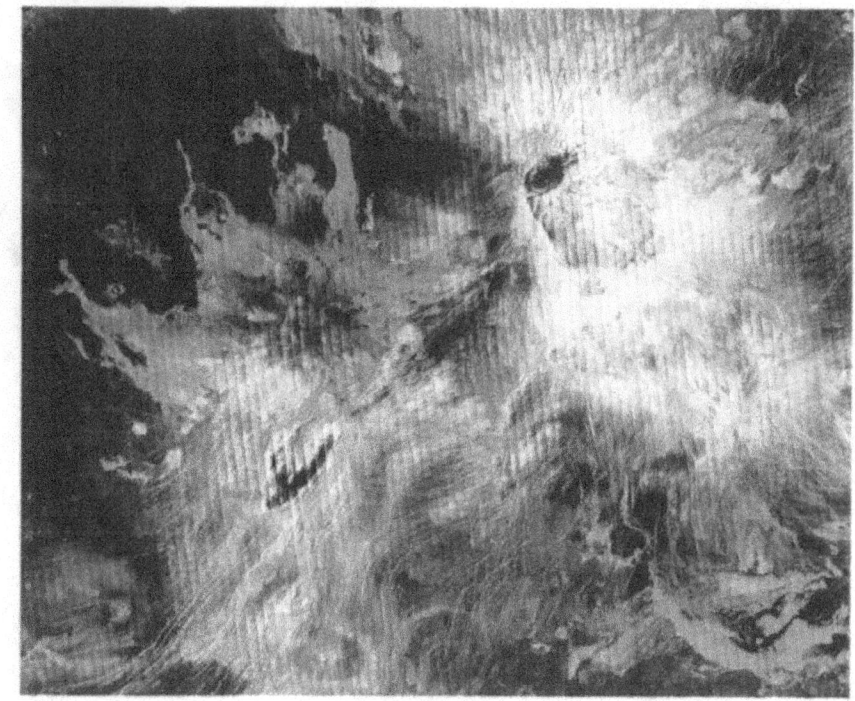

This 1500-km wide image shows Maat Mons (center left) and part of Ozza Mons (upper right). The numerous bright lines that cut through the center of Ozza Mons and the southeastern flanks of Maat Mons are faults. The alignment of many of these tectonic faults forms a rift zone. Ozza Mons is located at the junction of five rift zones, although only three (from the north, southwest, and southeast) can be seen on this image. At the lower right, exceptionally long flows originate from some of the fractures and continue for more than 1000 km (see Fig. 6.6). The association of rifting with volcanism is common on Venus; at Beta Regio a volcano (Theia Mons) is split in two by the rifting in Devana Chasma. **5.4**

ummit of Gula Mons (22.0°, 359.1°) in western Eistla Regio rises 3 km above the surrounding topography and is dominated by the linear bright structure seen in this image. The unusual summit comprises a series of fractures that links a large rift zone on the southeastern flanks of the volcano (not seen here) with the faults that extend to the north–northwest and surround the corona (top of image; see Fig. 5.17). Gula Mons consists of bright and dark radiating flows that reach up to 300 km from the summit. Flows from Gula's companion volcano Sif Mons (see Fig. 3.2) can be seen at the extreme left of this image. **5.5**

100 km

This unnamed volcano (–12.5°, 261.5°) is about 500 km in diameter and sits on top of an extensional fracture belt, but shows evidence of later fractures cutting the lava flows. This relationship suggests that volcanism and fracturing occurred at about the same time. The pear-shaped ridge feature at the center of the volcano is the outline of a central depression or caldera (see Fig. 5.8) that is associated with the volcanism. At the lower left, a small patch of tessera (high areas of extremely deformed terrain, see Chapter 8) has deflected lava flows from the volcano. **5.6**

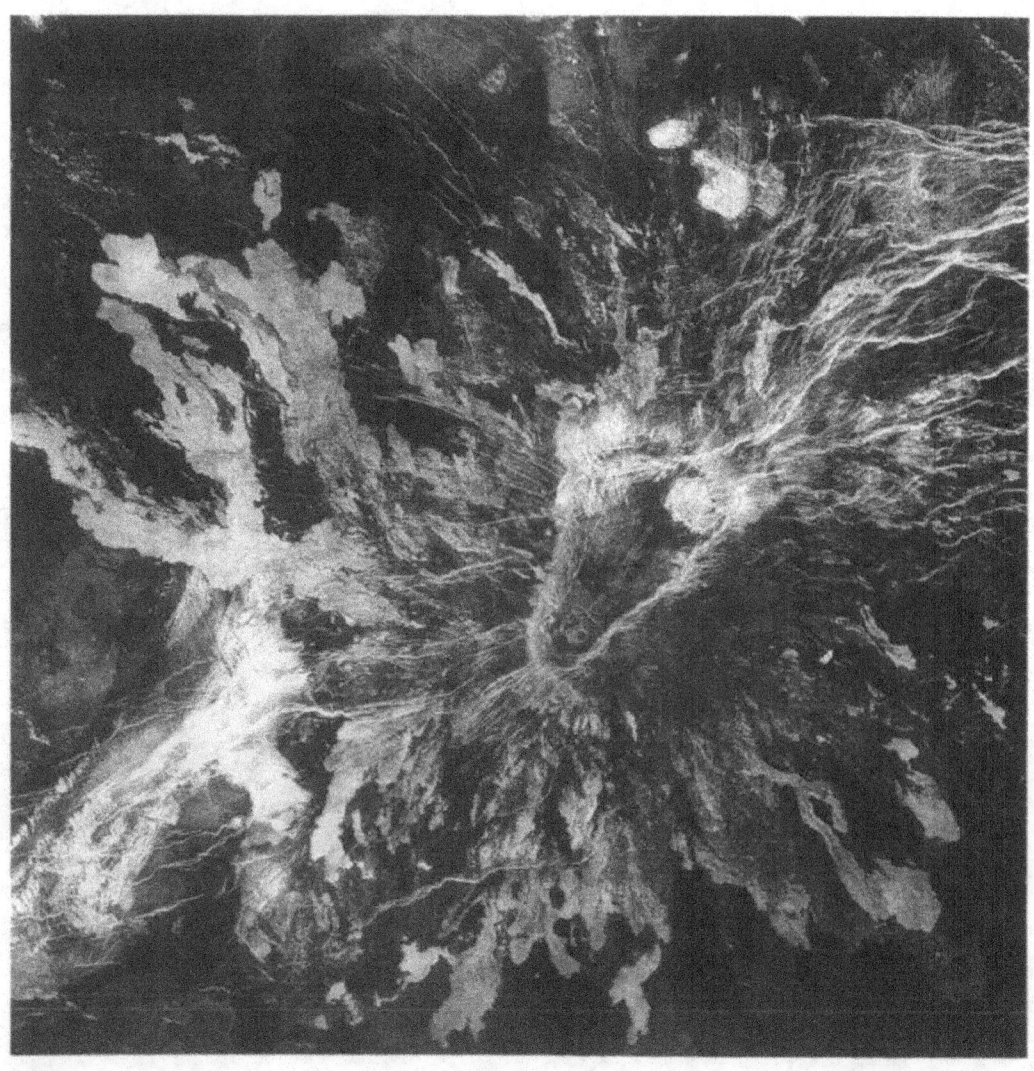

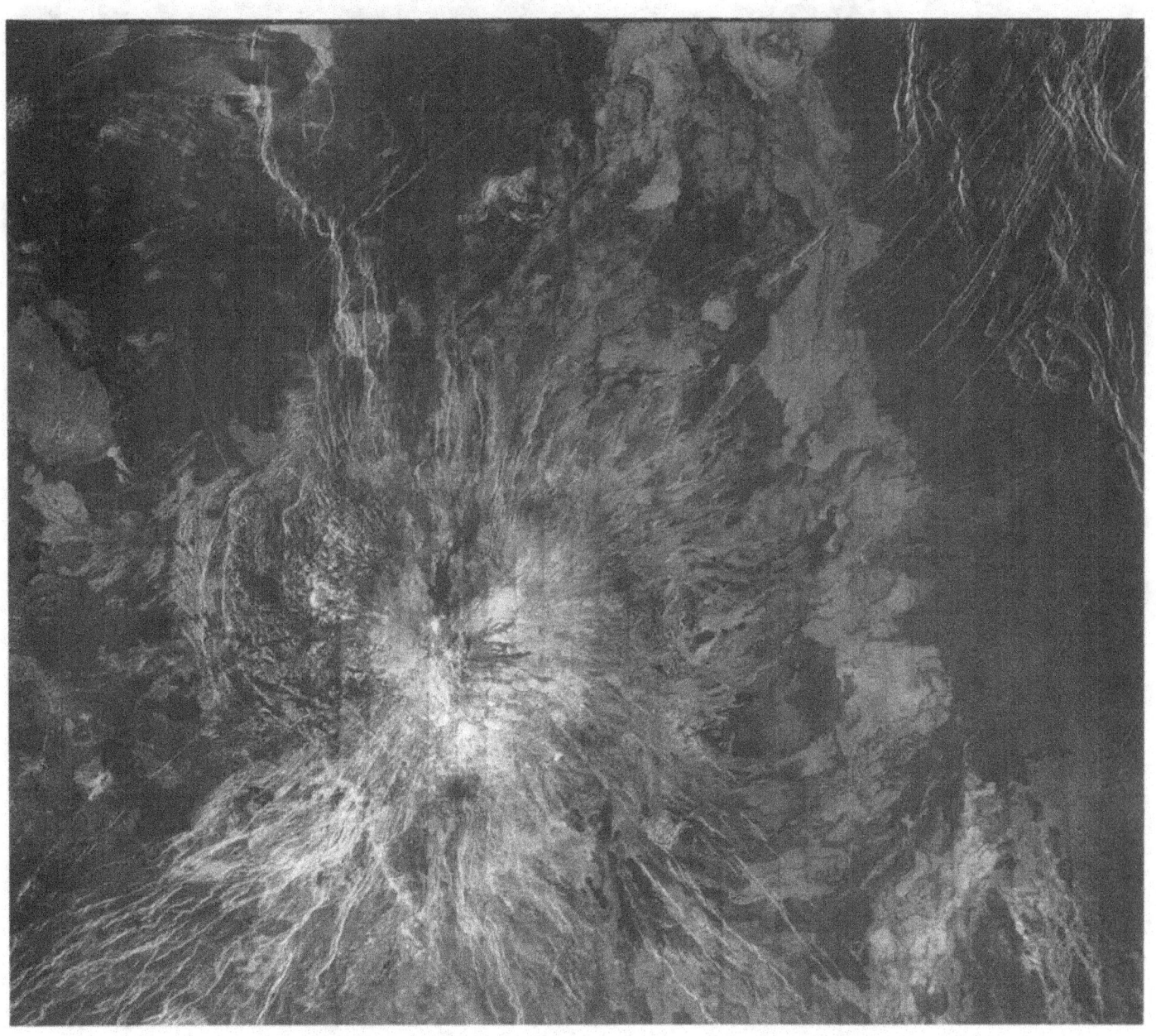

Approximately 80 percent of Venus' surface landforms can be attributed to volcanic processes

Located at (45.1°, 241.0°), this large volcano is 2.6 km high and has lava flows that reach up to 200 km from the center. The narrow flows near the summit broaden and pond as they reach the flatter ground to the east. On the western flanks (and on the east), tessera (see Chapter 8) have been partly flooded by the flows and a small shield. Radiating fractures extend out from near the summit for many kilometers. 5.7

50 km

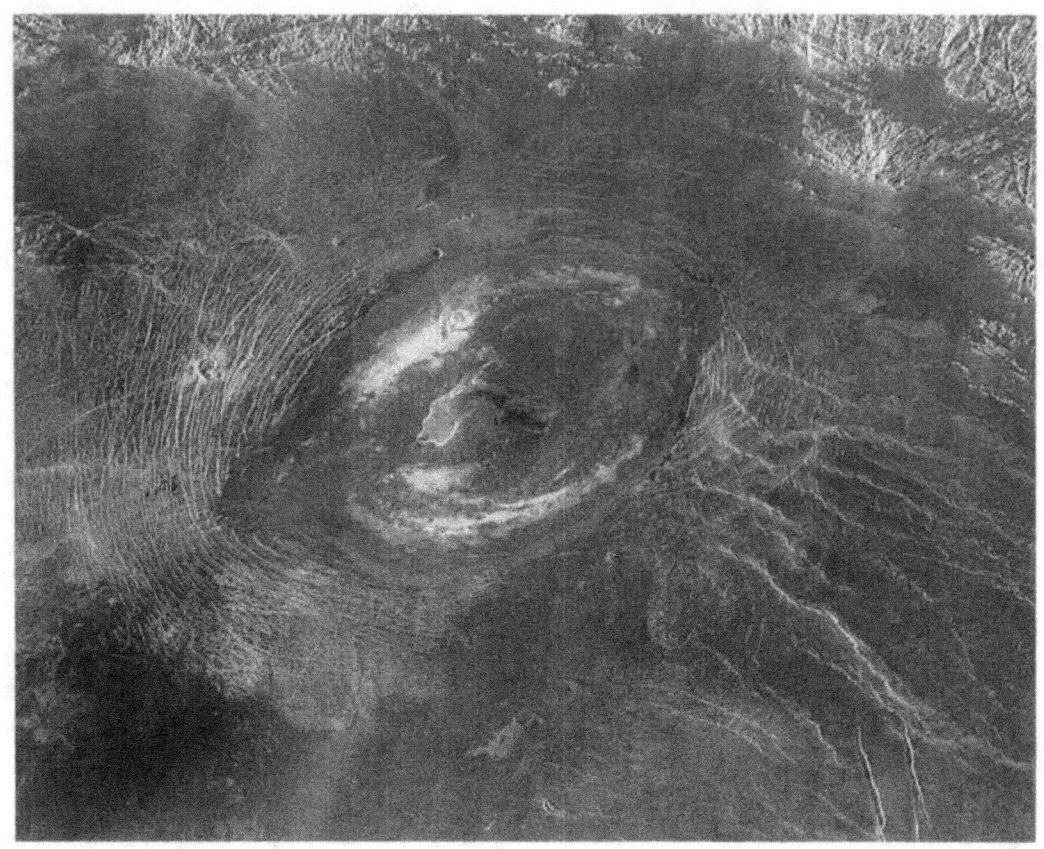

Sacajawea Caldera, located on the volcanic plains of Lakshmi Planum in Ishtar Terra, the northern highlands of Venus. Collapse calderas are circular to oval depressions, usually ringed or bounded by concentric fractures that surround a flat floor covered by volcanic flows. Although many calderas are located at the summit of large volcanoes, others are found on the plains, with flows that radiate away from their outside rims. In contrast to explosion calderas, which form when a volcano erupts violently and leaves a depression where the edifice once stood, collapse calderas form when shallow magma chambers deflate through eruption or transfer of magma elsewhere. This removes support for overlying rock, causing it to slump and form a broad, shallow depression. Both types of calderas occur on Earth, but Venus shows no evidence of explosion calderas while most of the collapse calderas occur on the plains. Sacajawea is 105 km by 150 km across and 1–2 km deep. Situated atop a broad region of somewhat raised topography within Lakshmi, the caldera is surrounded by smooth volcanic plains, while concentric fractures define its perimeter and bright patches inside indicate the youngest flows. To the north lie partially flooded tessera remnants. Tessera, highly deformed and ancient terrain, probably underlie Lakshmi's high plateau. **5.8**

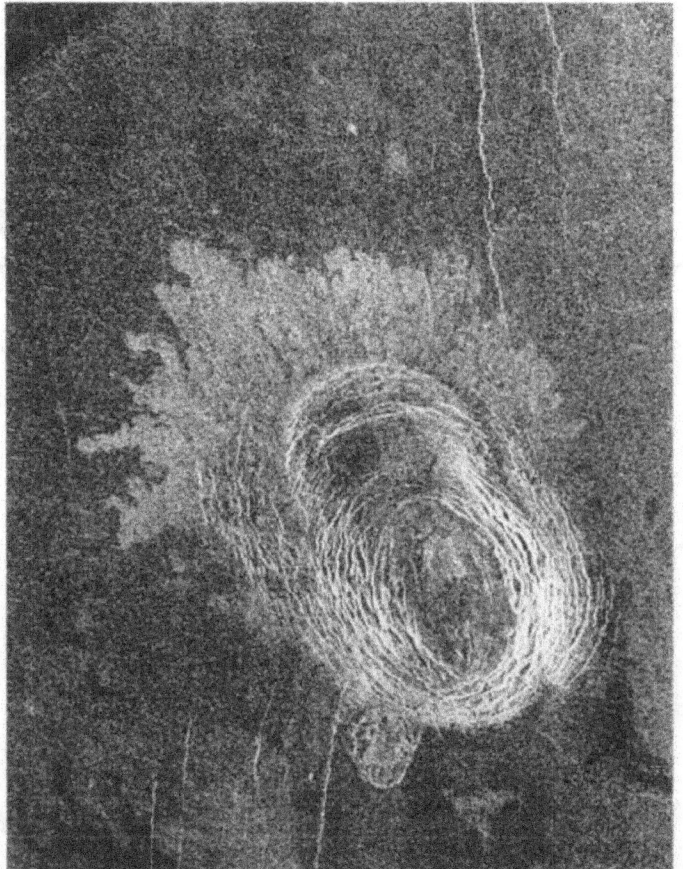

Sachs Patera is located in the plains south of Ishtar Terra in Venus' northern hemisphere. Sachs is defined as a "sag-caldera" since it lacks an associated edifice. Its flat floor is enclosed by a scarp ring, and volcanic flows radiate away from a section of its circumference. Sachs is 40 km by 25 km across, 130 m deep, and shows evidence of multiple phases of drainage and collapse. The caldera is located within mottled and ridged units of volcanic plains. **5.9**

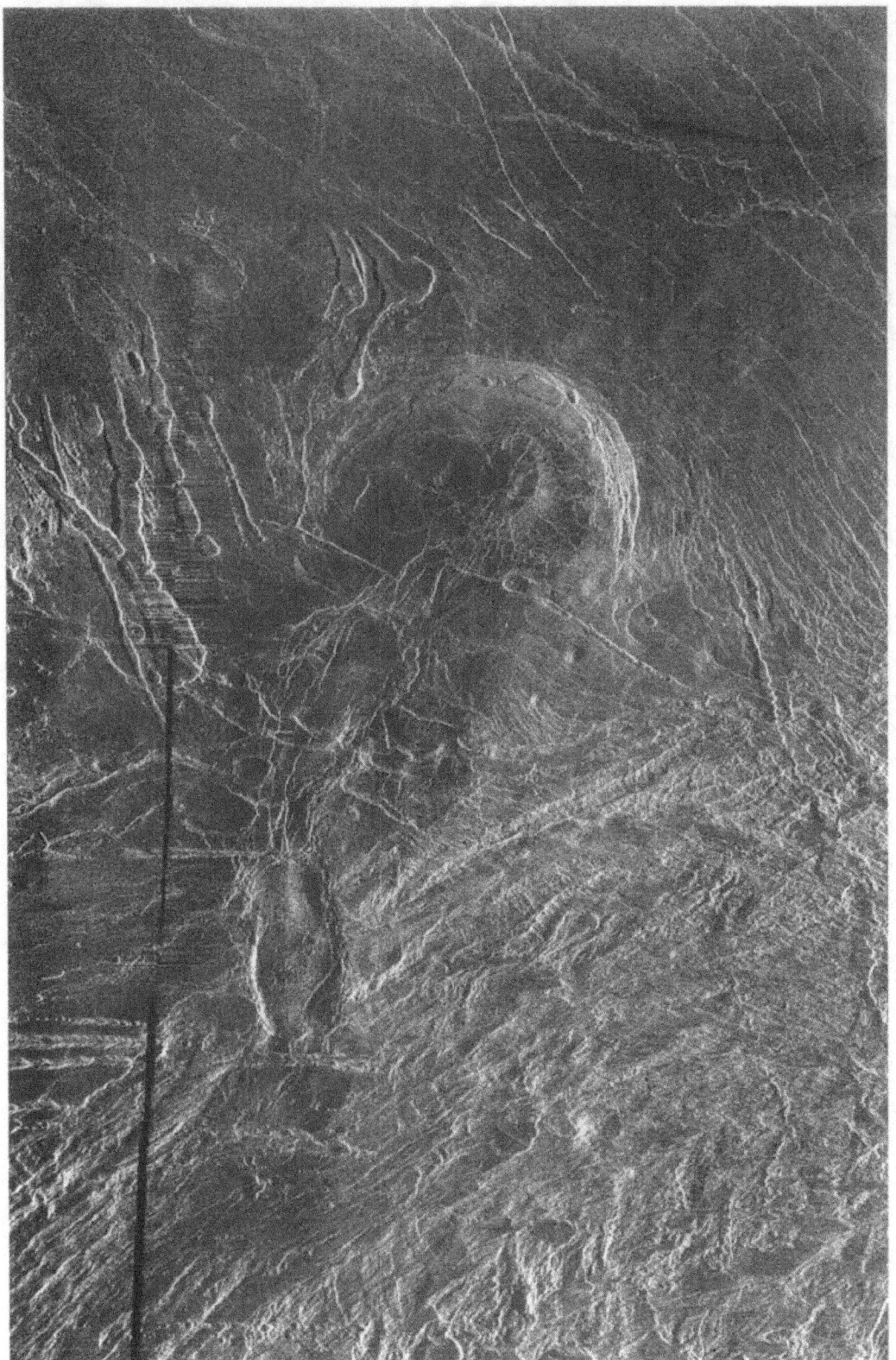

Siddons Caldera has a diameter of 64 km and lies at the southern edge of the smooth volcanic plains of Lakshmi Planum — where the plains meet Clotho Tessera, the highly deformed terrain to the south. The flat-floored and irregularly shaped troughs trending to the northwest may also be volcanic in origin. **5.10**

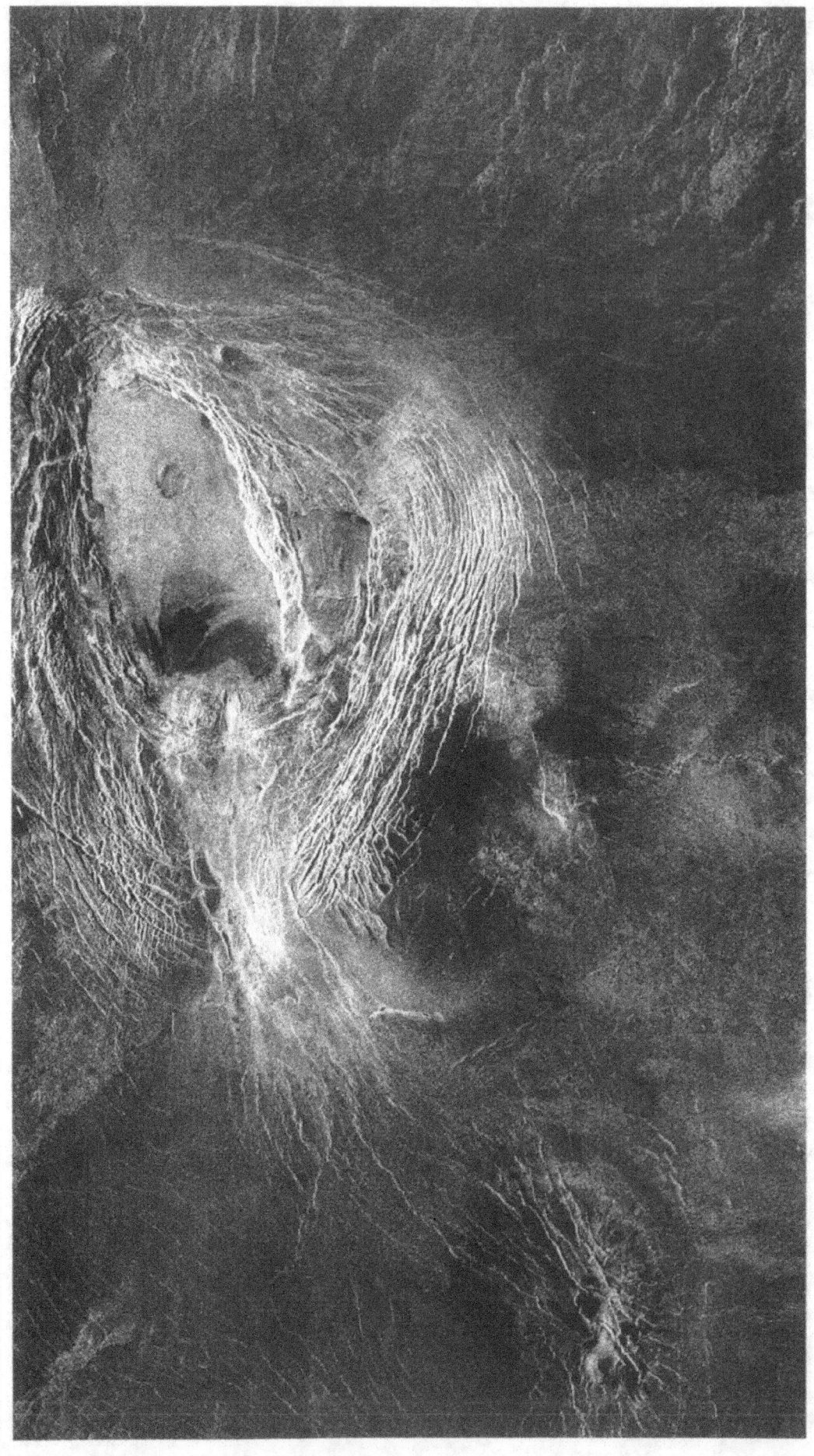

olette Caldera is located in the western part of Lakshmi Planum, atop a broad shield-like edifice more than 500 km across. This large caldera measures about 120 km by 65 km across and has a depth of about 2 km. The smooth volcanic floor is bounded by intensely fractured zones extending 50–80 km beyond the inner wall, including terraces between the outer fractures and the innermost scarp. A complex history and the effects of regional structures are shown in the noncircularity of the concentric fractures and by the gap in the caldera rim to the south. The mountains of Akna Montes lie just to the west of Colette and may have influenced its development. **5.11**

This circular caldera is located near the eastern end of Aphrodite Terra (6.0°, 227.0°), in the equatorial highlands of Venus. The radial lava flows and the broad band of concentric fractures and faults extend to a radius of nearly 60 km, while the inner scarp is about 15 km in radius, hinting that the zone influenced by the collapse is much wider than the caldera itself. Aphrodite Terra is a region of intense deformation and extension. This particular caldera is situated in a belt of fractures and graben within a region characterized by widespread and vigorous deformation and volcanism. **5.12**

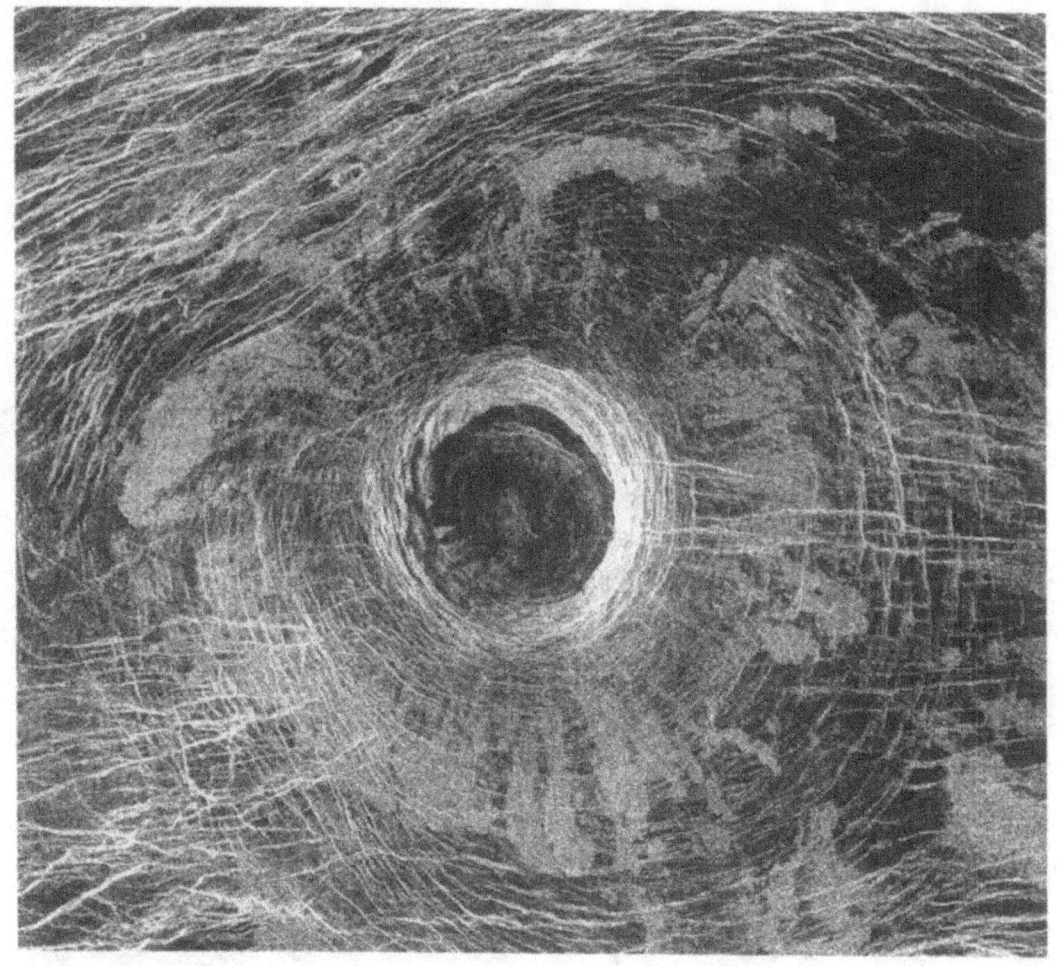

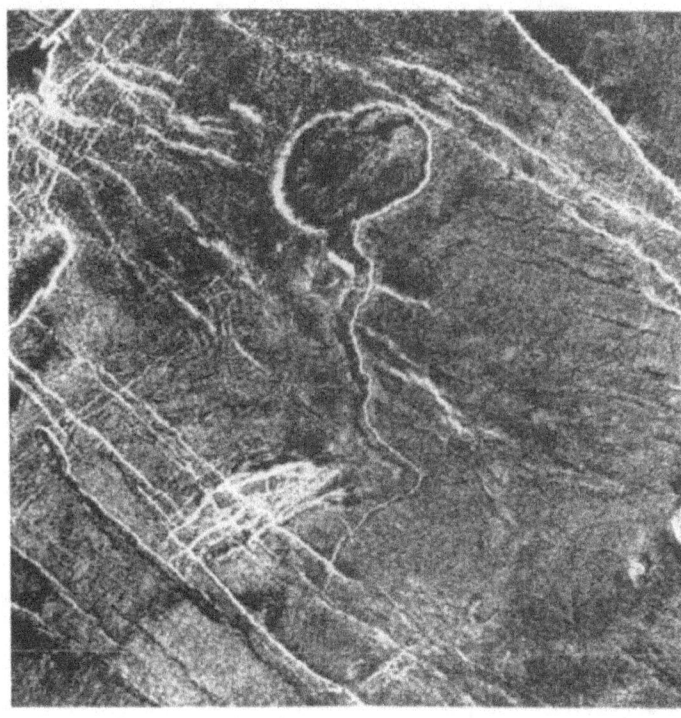

This small feature is located in eastern Aphrodite Terra (–25.5°, 197.5°) near a zone of ridged plains, dome fields, and coronae. Measuring 15 km by 10 km, it appears to have been formed by the collapse of a magma reservoir that is associated with the flow (which formed the channel or collapsed tunnel meandering toward the south). An alternative explanation is that the feature formed as a result of a very long duration lava eruption, which caused the underlying ground to thermally erode away and form a circular pit near the vent and a channel "downstream" (see Chapter 6). **5.13**

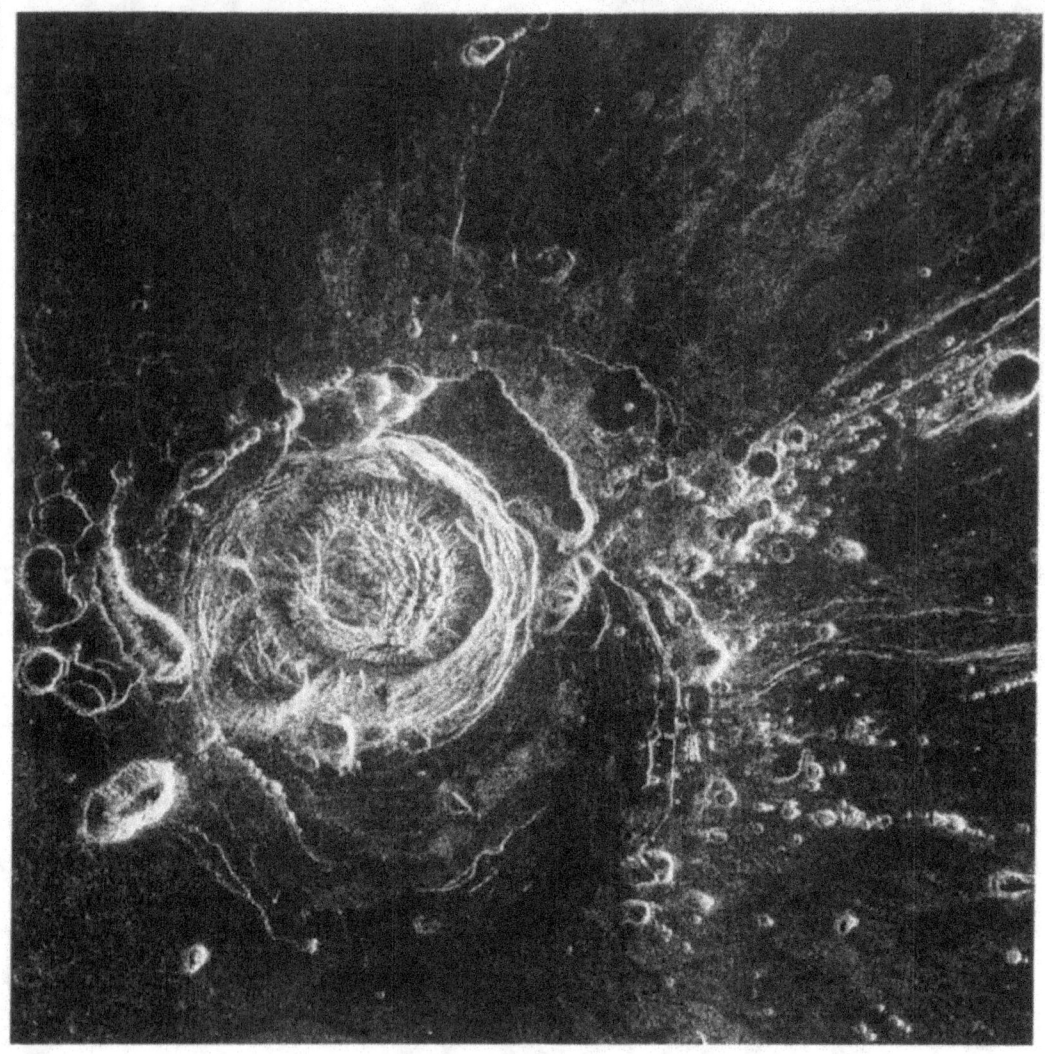

Numerous pits and pit chains surrounding this caldera, along with other lineations and extensive flows, imply a structural influence on the formation of this circular feature (9.0°, 29.0°). The central structure (about 20 km across) may be a caldera that formed over a magma reservoir from which numerous radial dikes branched. The dikes, connected to minor magma bodies, drained when the main chamber emptied, causing collapse along lines radiating away from the central caldera. Faint radial lava flows can be seen, particularly to the northwest. **5.14**

Venera images revealed a class of circular features, several hundred kilometers across, that were named "coronae" (Latin for "crowns"). The increased resolution of Magellan images of coronae revealed a huge diversity in form and complexity. The once-clear definition has become muddied: features described in later sections as arachnoids and novae could actually be considered related to coronae. In general, coronae are large, somewhat circular features with a tectonic annulus of ridges or grooves and some amount of associated volcanism. They are usually interpreted to be the sites of upwelling of hot mantle from deep in the interior of Venus. Nearly 400 features have been classified as coronae on Venus. This particular corona (–25.6°, 82.0°) is classified as concentric, is 350 km in diameter, has a 75-km wide annulus of fractures, and shows a moderate amount of associated volcanism. Note also the 40-km-wide dome on the western edge of the corona: this association of coronae with steep-sided domes is quite common. **5.15**

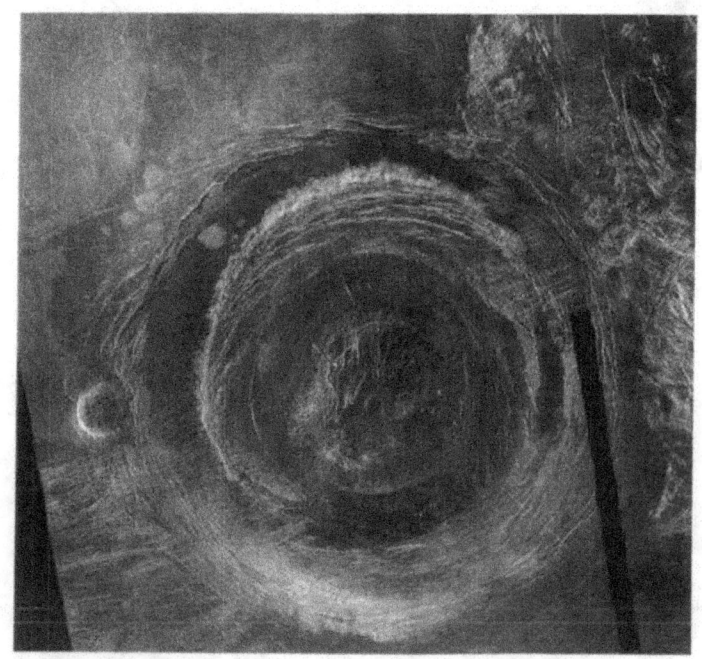

Heng-O Corona (2.0°, 355.0°). Although also classified as a concentric type, the 85-km-wide annulus is not as continuous or well-defined as that of the previous corona. Heng-O, which is 1060 km across, is the second largest corona on Venus and significantly larger than the average corona (275 km). Part of the eastern half of Heng-O was imaged from Earth by Goldstone radar and is shown in Fig. 1.8. The difference in viewing geometries between the two radar systems results in the different appearances of the corona in the two images. **5.16**

North of Gula Mons (Fig 5.5) are a pair of concentric coronae. The poorly defined western corona is approximately 300 km by 200 km and shows many faint, radial lava flows. To the east, a bright lava flow more than 200 km long has its source in the center of a 225-km-diameter corona. Other radial flows can also be seen breaching the 52-km-wide annulus of ridges and fractures. The fracture belt linking this corona with the summit of Gula Mons can be seen at the lower right of the image. **5.17**

This 310-km-diameter corona, located at (–16.5°, 17.2°), has a double ring that makes up the 85-km annulus of grooves. The corona also has some radiating fractures on its southern edge, as well as interior volcanism that has formed small shields. **5.18**

Several overlapping circular features make up this multiple corona (–16.4°, 347.5°), whose maximum width is 180 km. Many small volcanic shields are nearby and lava flows extend away to the west from the 20-km-wide annulus surrounding the largest part of the corona. The 15-km impact crater seems to cover some radiating fractures, suggesting that the crater formed after the corona. **5.19**

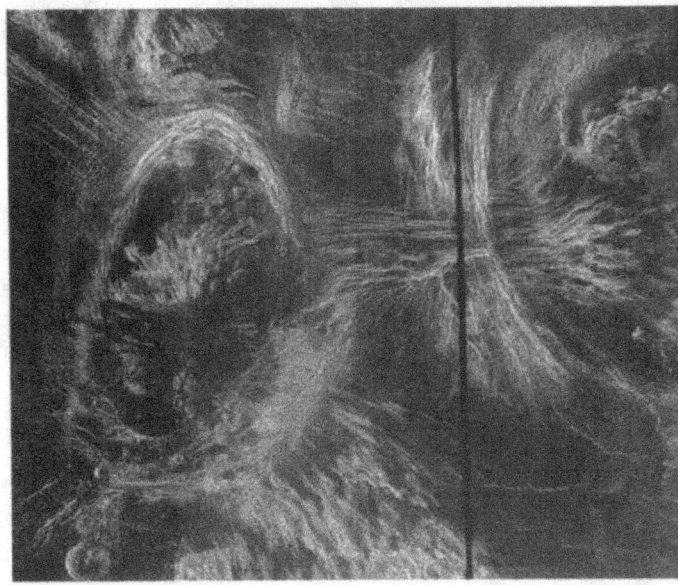

ocated about 1000 km south of Maxwell Montes at (49.0°, 2.0°), this pair of coronae are both approximately 300 km in length. The eastern corona, named Onatah, is asymmetric and has a broad annulus of fractures. The western corona is associated with many small domes and one larger steep-sided dome. The radiating fractures in the center are similar to structures elsewhere which, when better developed, are classified as novae (see images later in this chapter). 5.20

ocated at (−60.5°, 85.0°), this asymmetric corona is almost 700 km across. The smaller circular region on the southeast side is the source for numerous bright lava flows. The western annulus is partly buried by lava flows that may have originated at the many small cones in the center of the corona. 5.21

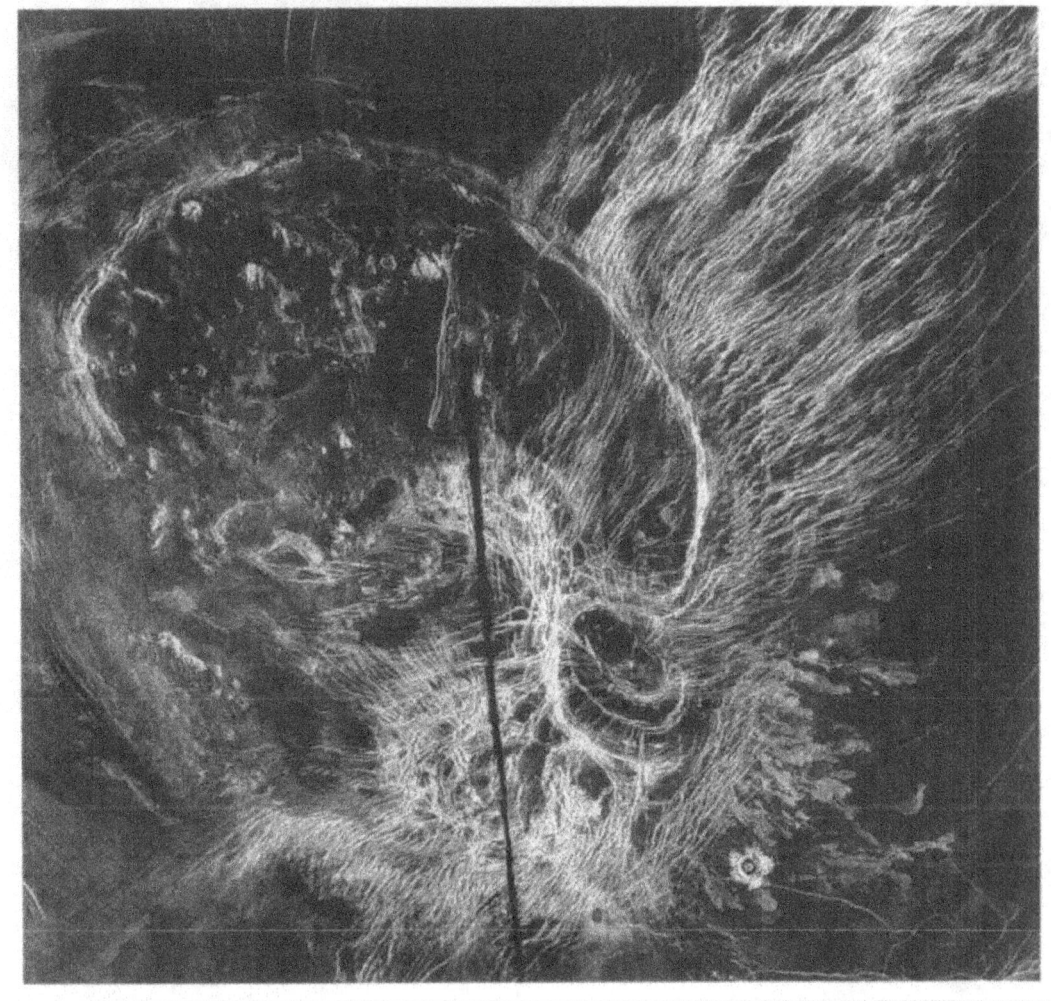

In addition to features such as volcanoes, which form primarily through the eruption of magma (i.e., through extrusion), Venus also has structures whose characteristics suggest that they formed when large amounts of magma approached near the surface of the planet without erupting (i.e., intrusive features). Such a process often produces a system of radar-bright lineaments at the surface that radiate away from a focal region. These radial geometries can originate as a response to doming of the surface when the presence of a large "diapir" (blob) of magma causes uplift through the injection of multiple narrow blades of magma (dikes) near the planet's surface, a process observed in Hawaii and elsewhere on Earth. The geometries may also result from the withdrawal of subsurface material and surface subsidence. If the lineaments are depressions (graben or fissures) in the surface and essentially radiate out from a central point, the structure is called a nova, a name indicative of its starburst-like shape. Approximately 50 such structures have been identified using the Magellan data set. Alternately, when the lineaments are either graben and fissures or ridges and extend several radii away from a central concentric structure, the feature is called an arachnoid, a name indicative of an overall spider-like appearance. At present, approximately 250 arachnoids have been observed on Venus. The starburst pattern characteristic of a nova is readily seen in the first example. This image, centered at (27.0°, 343.0°), shows that the nova has been superimposed upon a complex background of volcanic plains, but has also been partially buried by younger lavas, seen best in the northwest corner.
5.22

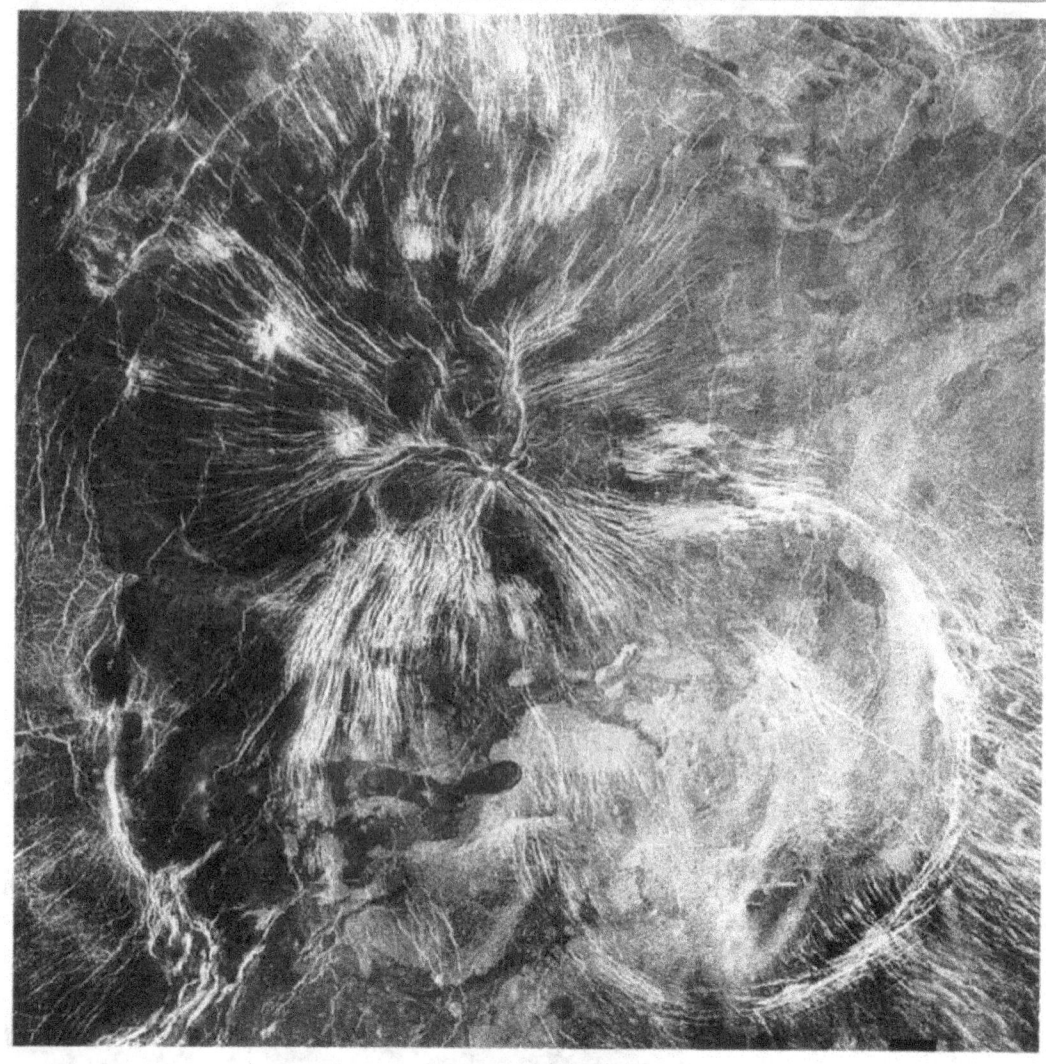

——— 40 km

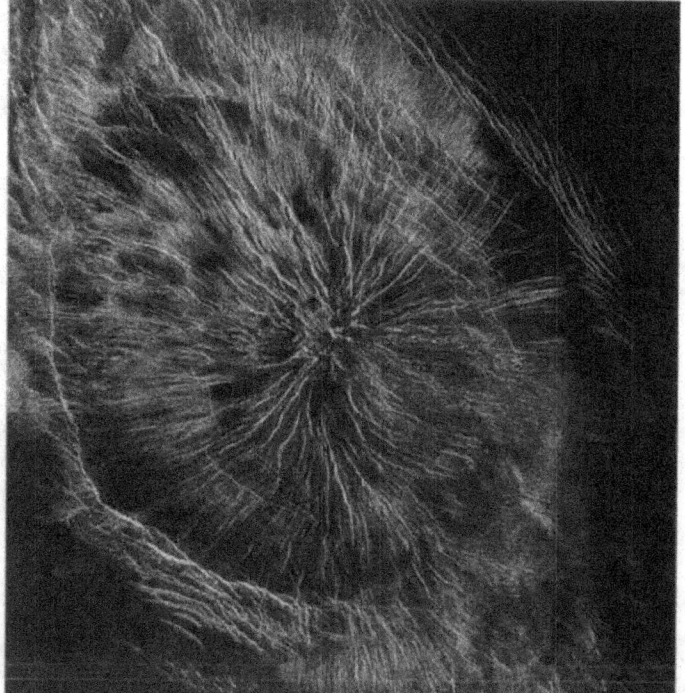

Unlike Fig. 5.22, this nova is contained within a partially developed ring of radar-bright lineaments, a configuration particularly indicative of formation through uplift. Within the ring there is little evidence of volcanic products associated with the radial lineaments; however, beyond the ring, long narrow flows can (barely) be seen extending away from the tips of the starburst pattern. The image is centered at (8.0°, 243.0°).
5.23

——— 40 km

As in Fig. 5.23, this nova is centered within a partially developed ring of lineaments. However, in this instance the radial pattern extends well beyond the confines of the ring (several hundreds of kilometers further than shown to the northwest and southeast), behavior more consistent with formation through dike injection. The image is centered at (–42.5°, 6.0°). **5.24**

————————————— 90 km

This image, centered at (41.5°, 214.0°), shows several arachnoids. As seen most clearly in the southwest corner of the image, these structures are often less radial than the novae shown previously, and the radiating ridges and central ring are superimposed upon a complex, varied system of volcanic plains material. Such radially ridged structures may indicate subsidence of the surface triggered by the withdrawal of material from a large subsurface magma body. **5.25**

————————————— 70 km

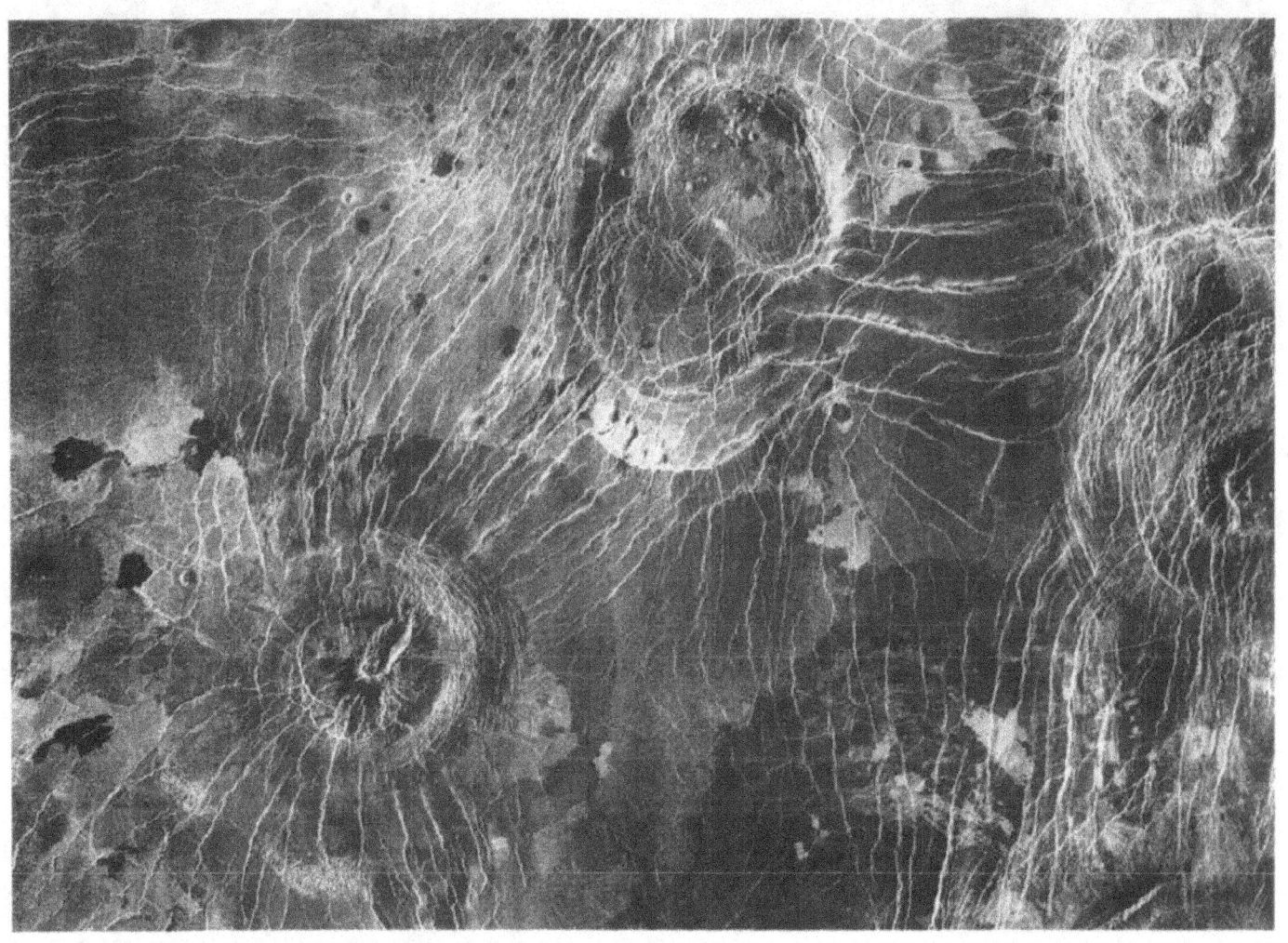

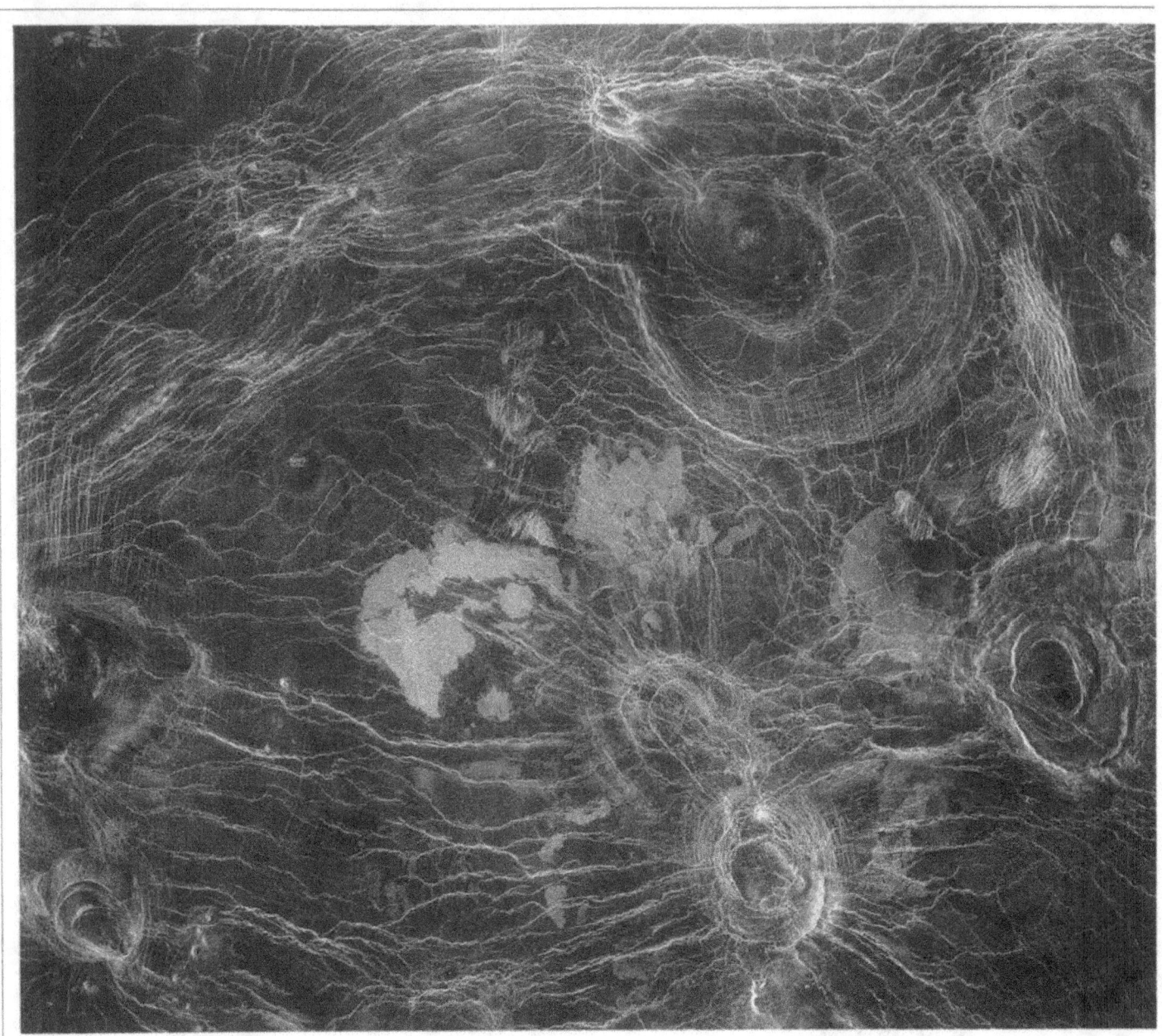

A s in Fig. 5.25, this region exhibits a high density of arachnoids
defined by radiating systems of
ridges. The arachnoids themselves are
of variable sizes, ranging from approximately 50–250 km in diameter, and
are generally superimposed upon the
background volcanism — as is illus-
trated by the ridges that crosscut the
radar-bright volcanic flows near the
center of the image. The area shown
is centered at (40.0°, 18.0°). 5.26

Some 250 spidery "arachnoid" features
have been observed on the surface of Venus

70 km

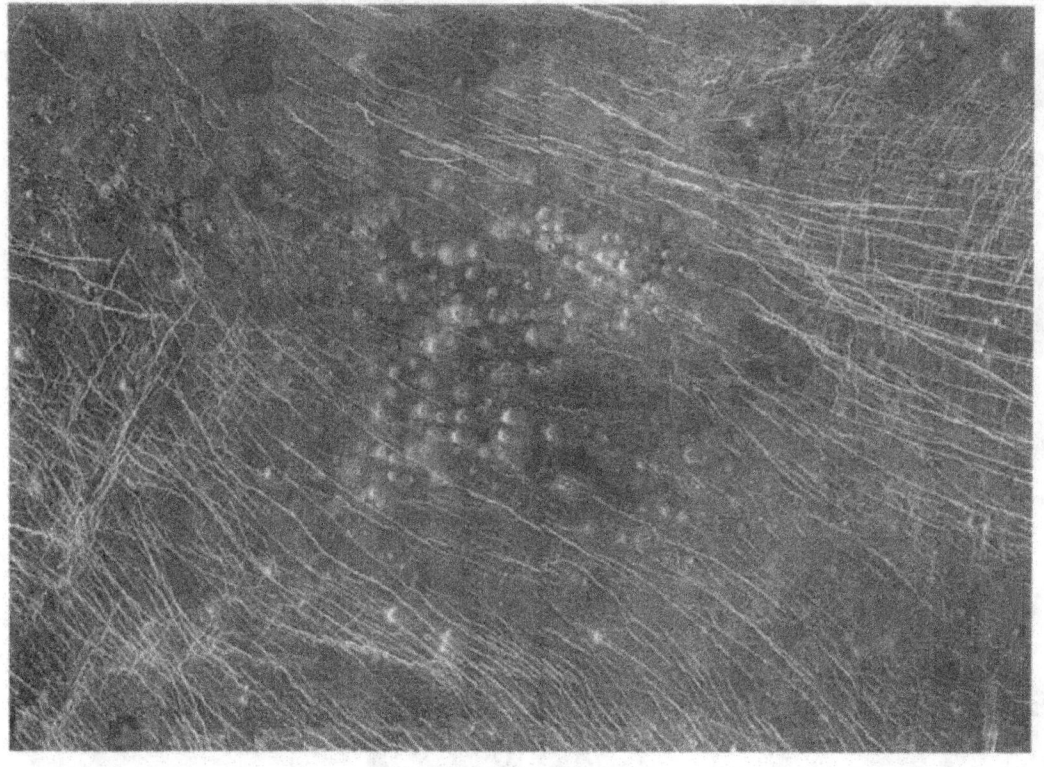

Planetary geologists are accustomed to working with large-scale geological features on most planets. Venus, however, because of the high-resolution Magellan data set, exhibits a large number of relatively small geological features that can be examined and that may prove to be important in understanding the geological evolution of the planet. Small volcanoes, less than 20 km in diameter, are extremely abundant on Venus and may number globally in the hundreds of thousands to millions. These volcanoes were first recognized in the Venera 15/16 and Earth-based radar images and were originally called "small domes" — using the lunar volcanic nomenclature. These domes are now interpreted to be predominantly "shield" volcanoes, similar to those forming the Hawaiian islands and composed of many small individual lava flows, although a few may be completely or partially formed by more explosive volcanic processes or single eruption events. The volcanoes are generally 1–15 km in diameter and less than 1 km in height. They frequently have summit craters but only rarely show visible individual lava flows; they also tend to occur in clusters of tens to hundreds of volcanoes called "shield fields," which are most commonly found on the plains of Venus and are also frequently found in association with larger volcanic or presumed volcanic features such as large volcanoes and coronae. This image shows a good example of a typical shield field. The field is centered at (34.8°, 331.3°) and covers an area approximately 80 km by 100 km. 5.28

Unlike Figs. 5.25 and 5.26, the arachnoid shown here is composed primarily of radiating graben and fissures rather than ridges. There are numerous small domes in the center of the structure, and portions of the elliptical outer ring have been flooded by radar-dark lava. The radiating lineaments to the south, and thus presumably those to the north as well, have been emplaced in at least two separate events. A dark lava flow in the southern portion of the image (unrelated to the arachnoid) has buried many lineaments, but one set clearly cuts across the flow, indicating a younger age and sequential development of the radiating feature. The area shown is centered at (-16.5°, 352.0°). 5.27

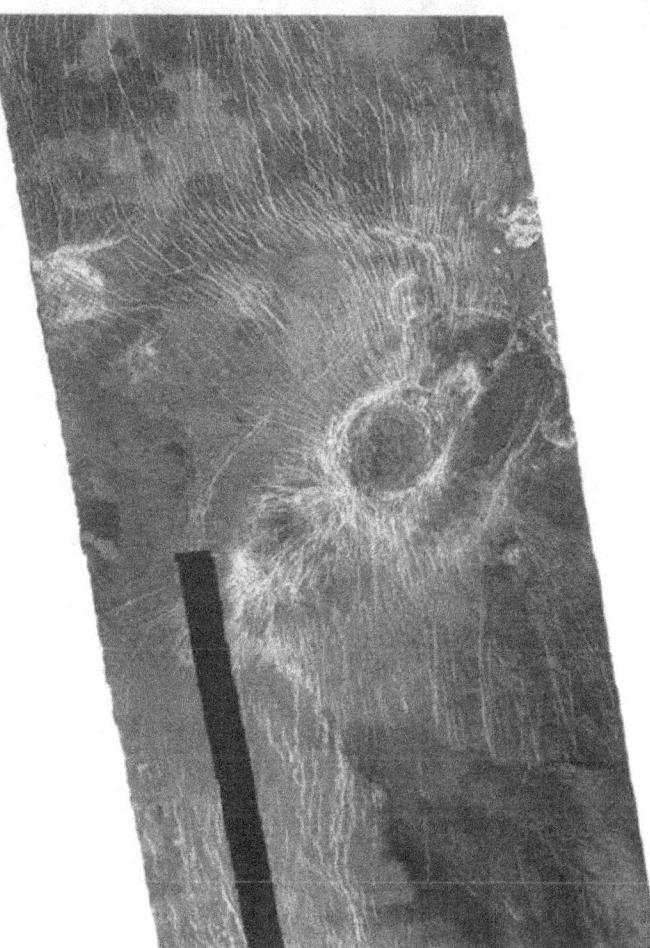

▬▬▬▬▬▬ 80 km

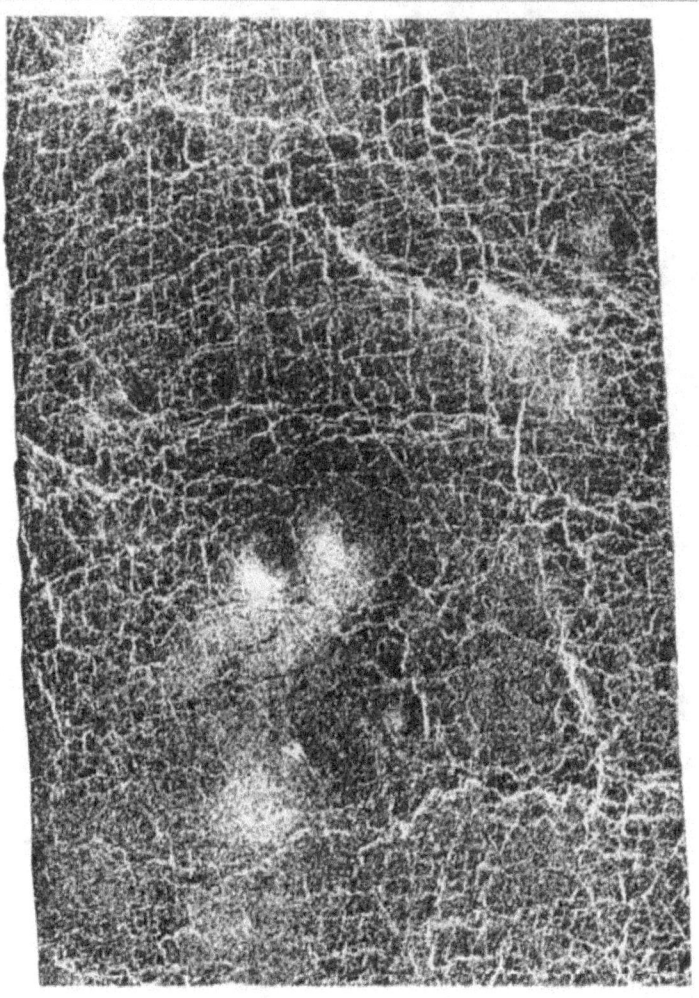

The small volcanoes of Venus occur in a variety of morphological types. These images show two examples at the same scale. Each image covers 75 km north–south and 45 km east–west. Above (left) is a small cluster (part of a larger field) of relatively steep-sided volcanoes, ranging in diameter from 2 km to 5 km, that is visible as radar-bright and radar-dark patterns related to topographic slope. Some of the volcanoes appear to have summit craters but most do not. The small volcanoes appear to postdate and cover the bright lineament patterns in the plains that are interpreted to represent fractures. Radar-bright and radar-dark streaks and patterns around the volcanoes may represent either material erupted from the volcanoes or wind streaks caused by removal of previously deposited material. This cluster is centered at (25.0°, 330.6)°. Above (right) is a cluster of volcanoes (part of a larger field), ranging in diameter from 2 km to 15 km, that appears to be defined solely on the basis of radar "darkness." Although many have central craters, these volcanoes do not appear to show any topographic slopes and are much more irregular in outline than the volcanoes in the image at left. These small volcanoes may be very low and broad in profile and produced by radar-dark lava flows that, in some cases, can be seen to flow into and along fractures in the plains. This cluster is centered at (–32.5°, 337.0°), south of the "crater farm" (Fig. 4.36). A third morphology is seen in other small shields and is repeated in the intermediate-size (10–100-km diameter) volcanoes (Fig. 3.8). These volcanoes comprise very distinct, finger-like lava flows of uniform length that tend to brighten away from the summit. Most of these volcanoes have summit collapse features. 5.29

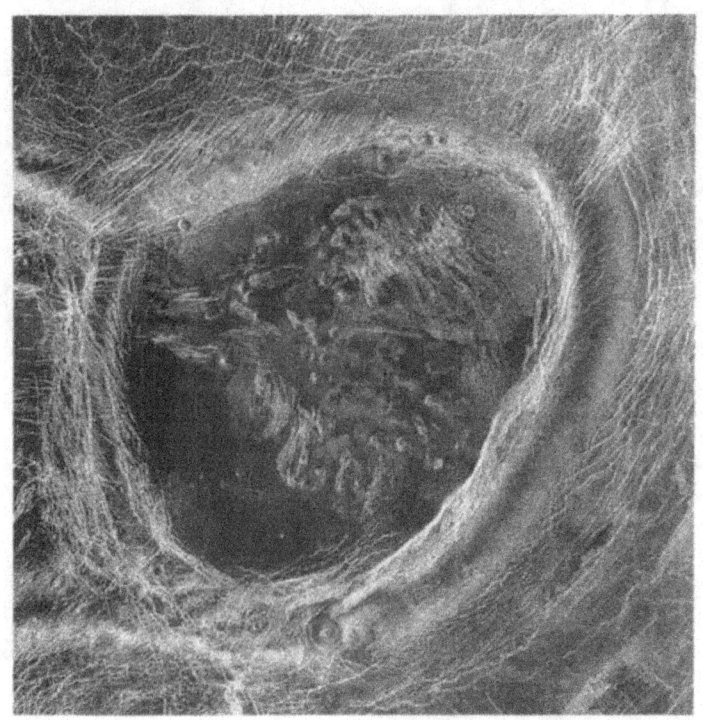

hield fields frequently occur in association with coronae, generally inside the arcuate corona structure. Associated shield fields can predate, postdate, or form contemporaneously with, the development of an associated corona. In this example (50.0°, 205.0°), the small shield field consisting of volcanoes ranging in diameter from 1 km to 9 km is located within the inner ring of a corona (Neyterkob) that is approximately 200 km in diameter. The stratigraphy of the mappable plains units within the center of the corona and the occurrence of small volcanoes on the peripheral structure of the corona indicate that this shield field is either contemporaneous with or postdates the development of the corona. **5.30**

hield fields also commonly occur in association with large volcanoes, frequently near the summit or on the flanks. This example (4.0°, 200.0°) shows a field of several hundred volcanoes, predominantly 5 km and less in diameter, that occurs on the north flank near the summit of the large (1000-km-diameter) volcano Ozza Mons. The shield field is approximately 80 km in diameter; volcanic material associated with the small volcanoes has buried structural grooves and ridges in this area. **5.31**

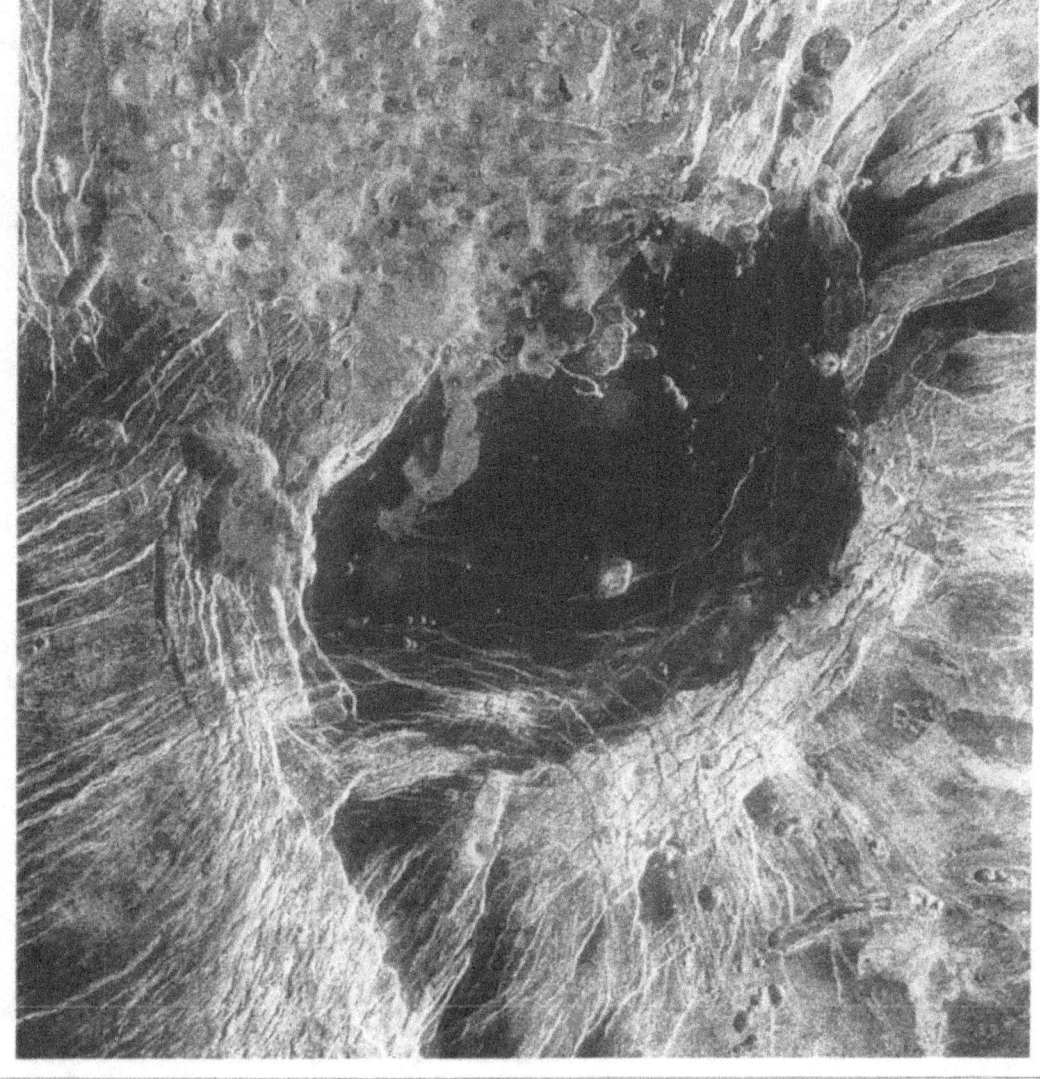

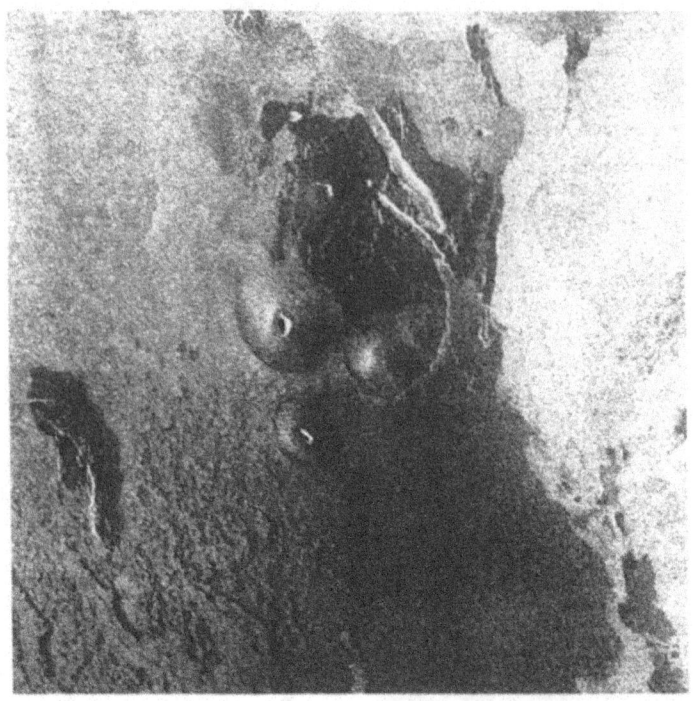

In addition to the shield fields occurring on the plains and associated with large volcanoes, on many occasions only a few small shields are found. In this example, three radar-dark and well-defined shields, each about 5–7 km across (two have pits), are located on the northern flanks of Maat Mons (see Figs. 5.1 and 5.4). These shields have been partly surrounded by dark deposits from the south. Just north of these are several small pits likely associated with other shields. The bright lava flow on the right (from Maat Mons to the south) has deflected around one of these shields, defined by its moderate brightness and central pit. The width of this entire image is about 80 km, only a fraction of the size of Maat Mons. 5.32

Four domes (25–50 km across) lie clustered in the lowlands of Guinevere Planitia (34.0°, 312.0°). The scalloped edge of the northernmost dome is probably the result of material that has slid away from the dome and been subsequently covered by lava flows. The southeastern dome may have suffered this slumping process multiple times. As in the previous example, the circular growth of later-formed domes is hindered by the presence of adjacent domes. Fractures from the south cut these domes, suggesting that the domes are relatively old. 5.33

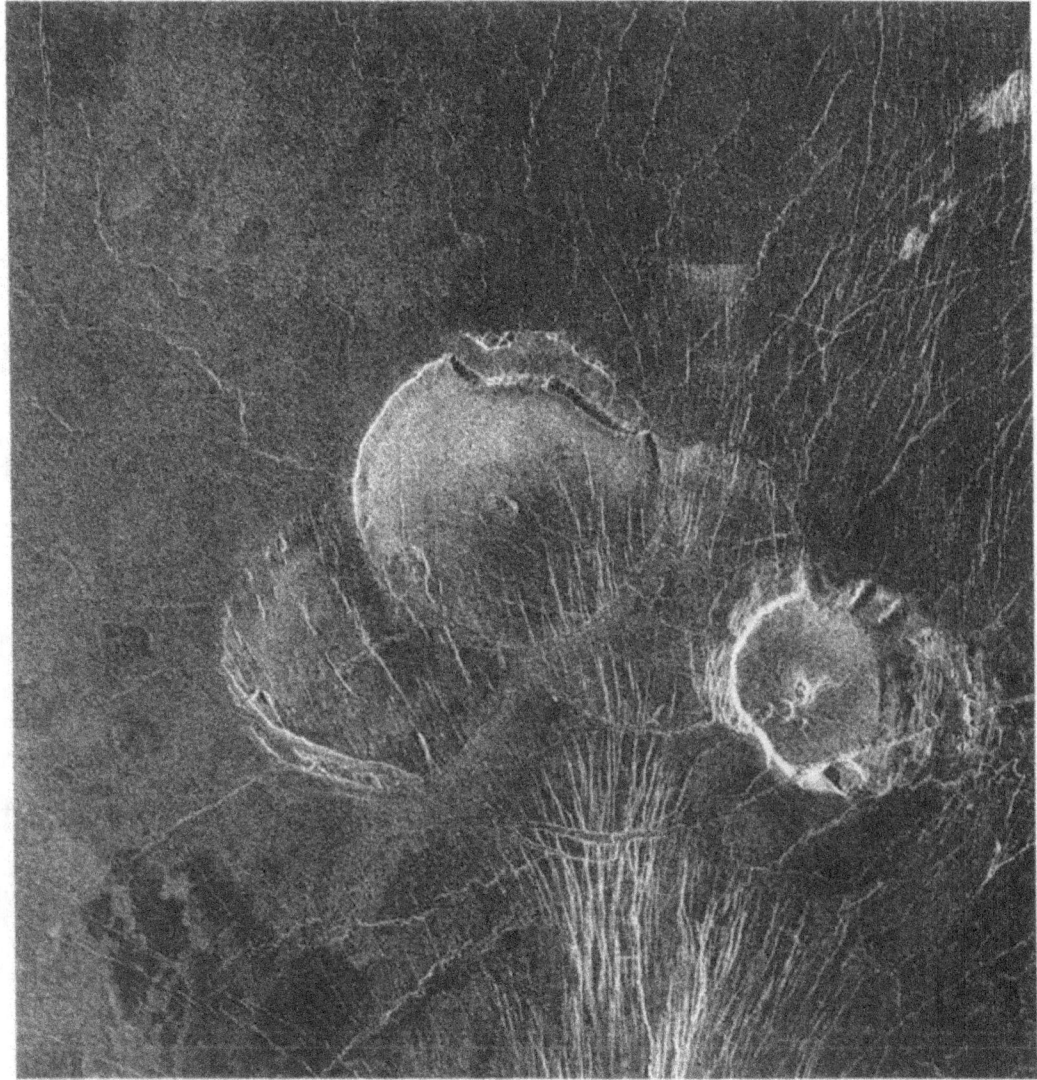

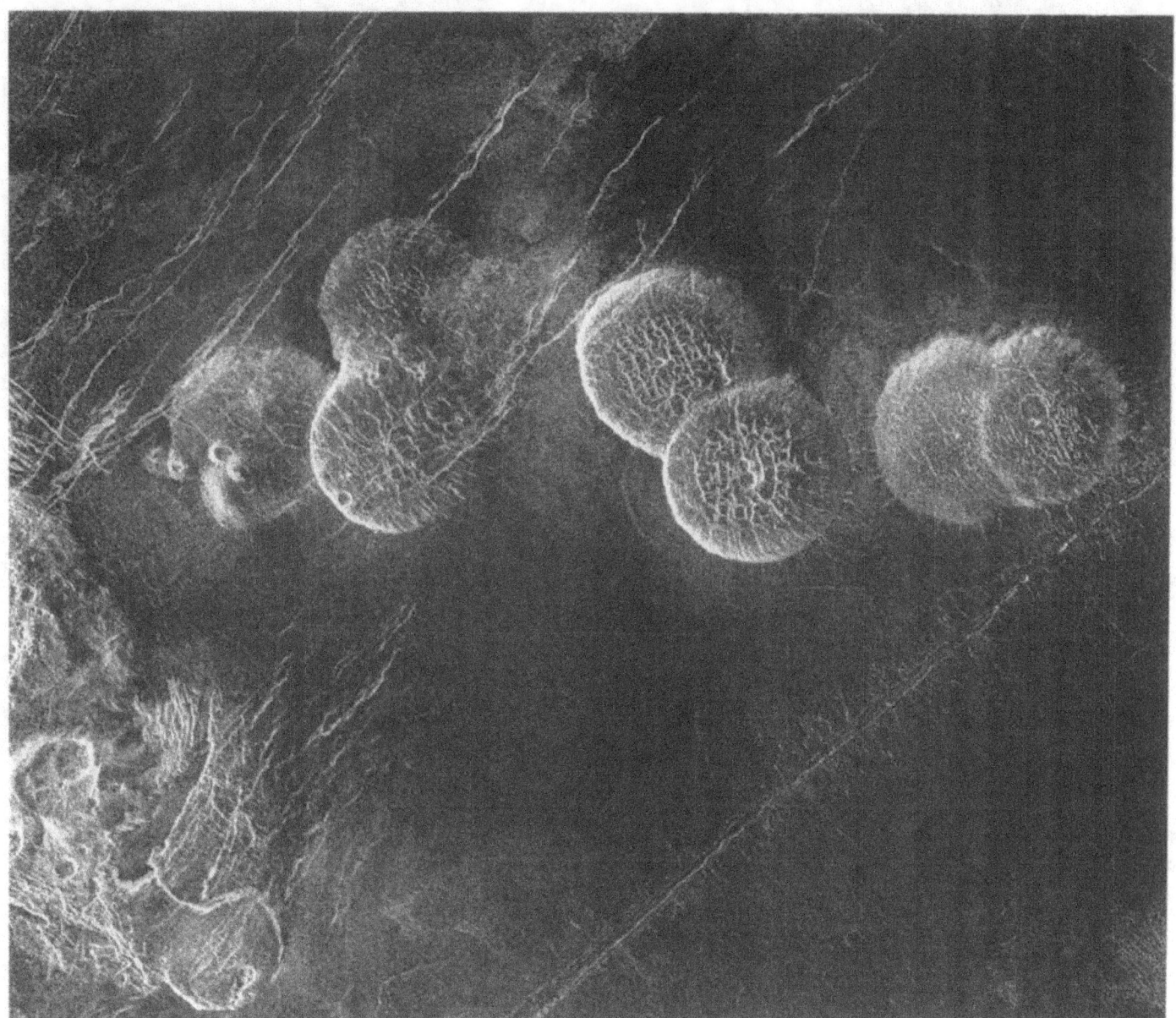

One of the first images returned by Magellan shows this area centered at (−30.0°, 11.8°) and lying to the east of Alpha Regio. Seven domical hills are displayed, averaging 25 km in diameter and rising an average of 750 m above the surrounding plains. These features are characterized by steep sides and flat tops (hence their informal name — "pancake domes") and are morphologically similar to several domes on Earth composed of viscous silica-rich lavas. The venusian domes, however, are 10–100 times wider than their terrestrial counterparts. The domes are interpreted as having been formed by continuous eruptions of viscous (sticky) lavas onto relatively level ground. The concentric and radial fractures associated with the domes may be produced by the expansion of a solidified crust or subsidence from cooling or magma withdrawal. The radar-bright edges of the domes may be indicative of talus or debris associated with mass wasting processes. Steep-sided domes are widely distributed over Venus, but are generally associated with coronae and tesserae. These particular domes are significant in that they are the primary evidence for lavas of evolved chemical composition on Venus. 5.34

Volcanoes on Venus are generally much wider and somewhat shorter than those on Earth

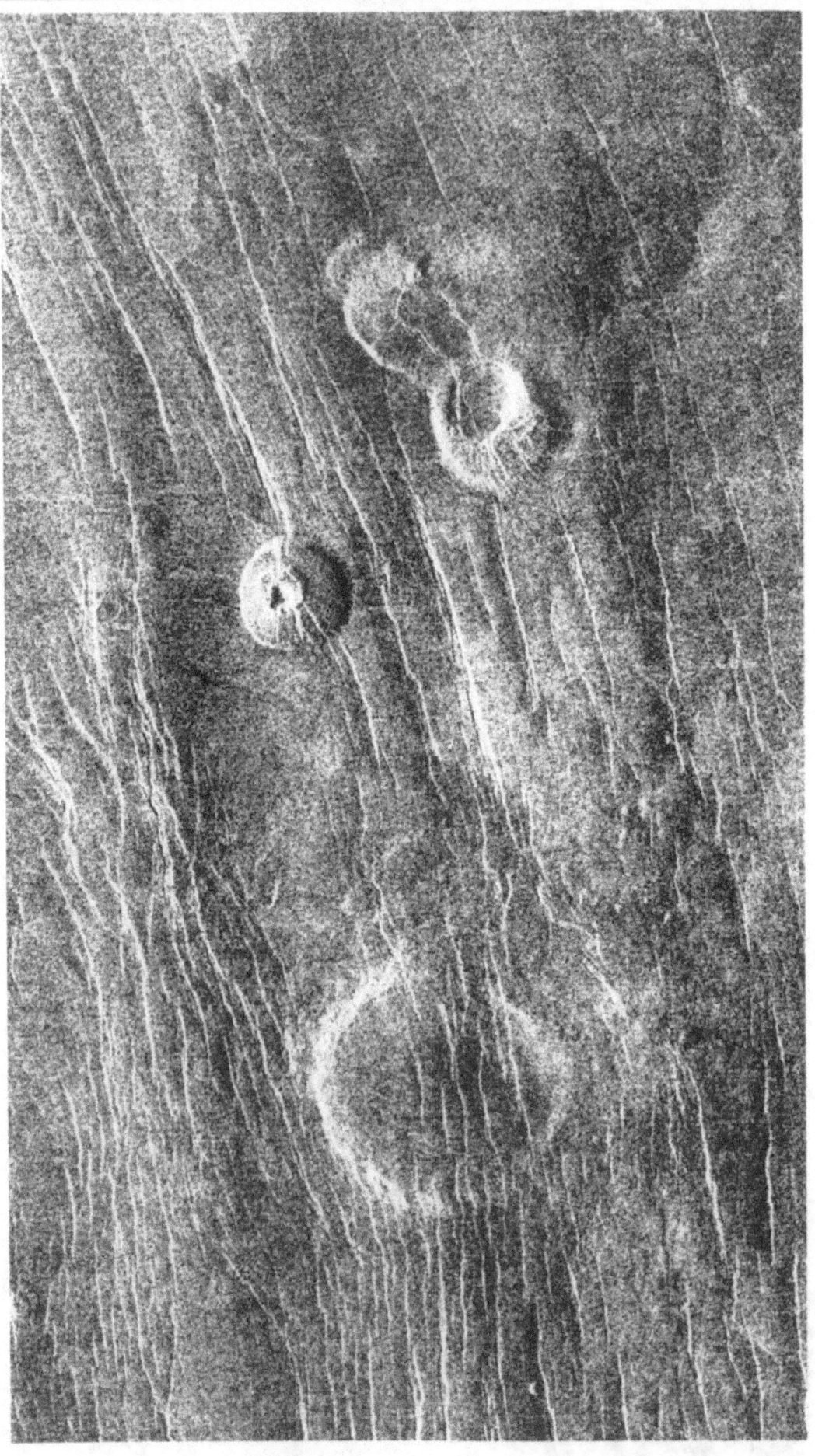

Domes on Venus come in a variety of shapes. In addition to the flat-topped pancake domes, there are also domes with rounded tops, sometimes with central collapse pits. This image (–6.0°, 333.0°) shows three domes of different morphologies, all cut by fractures. At the bottom of the image is a faint pancake dome (20 km across). At the top of the image are two round-topped domes (approximately 15 km across). The southern dome is very round and has a summit pit. The northern dome has collapsed on the northern margin, leaving a large scalloped edge. The rounded outline of the collapsed material suggests that it may still have been somewhat molten when the collapse occurred. **5.35**

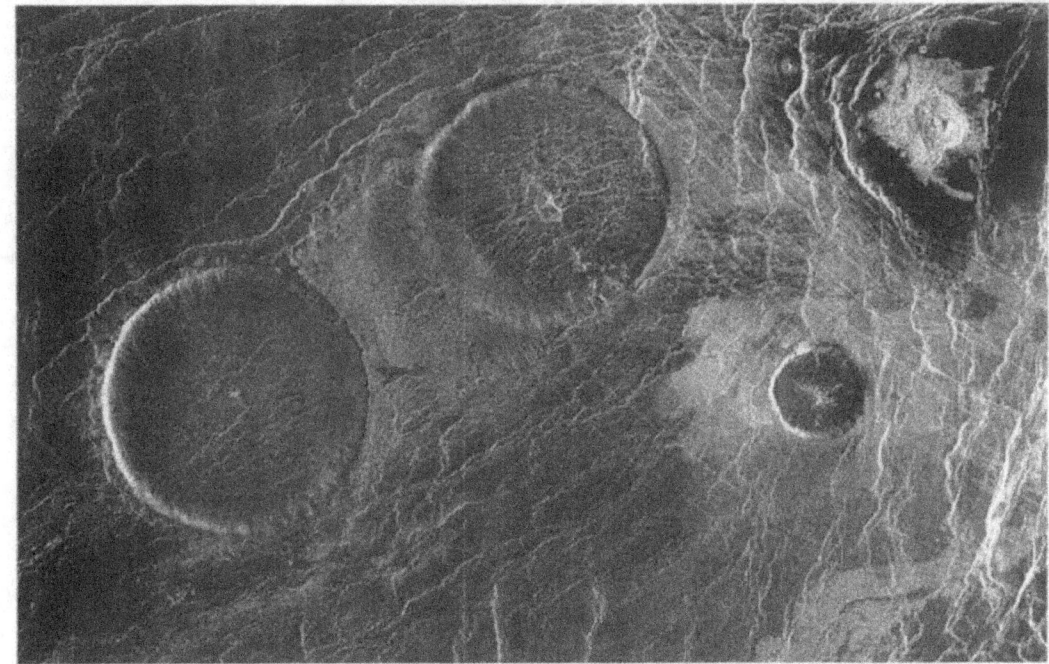

These three domes are located in Tinatin Planitia (12.0°, 8.0°), on plains where compression has formed sinuous ridges. The westernmost dome is 62 km in diameter. The western edge of the northernmost dome has slumped, releasing volcanic material (slightly lighter, smoother material) that flowed southwest and embayed the neighboring dome. This breakout process also occurs in terrestrial domes. **5.36**

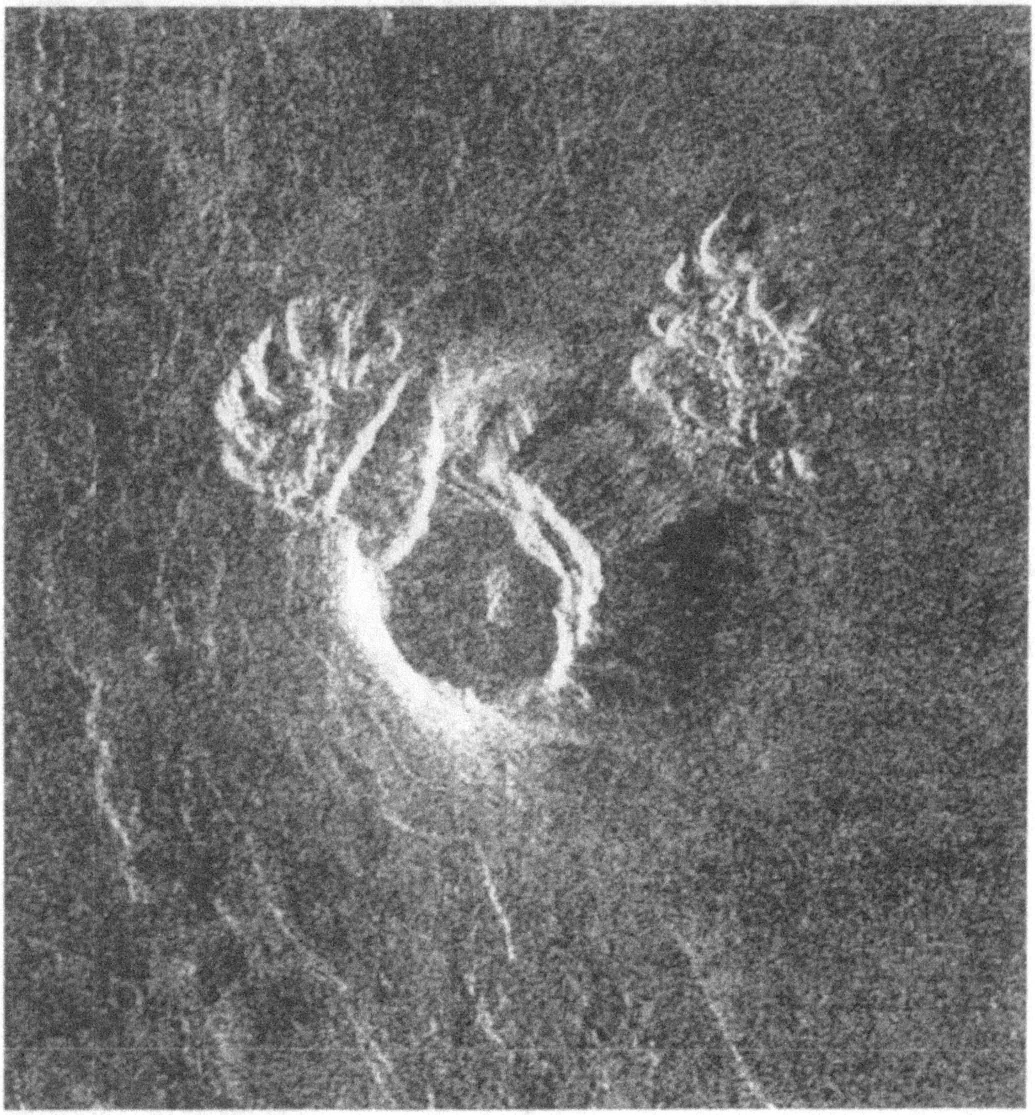

In contrast to the fluid collapse shown in Fig. 5.35, the irregular and rubbly appearance of this small dome suggests that it experienced two collapses once the material solidified. This dome, located in Navka Planitia, at (−25.0°, 308.0°) is approximately 17 km across and almost 2 km high. The dimensions result in a steep slope (about 23 deg), significantly steeper than the 1-deg slopes common on Venus' large volcanoes. **5.37**

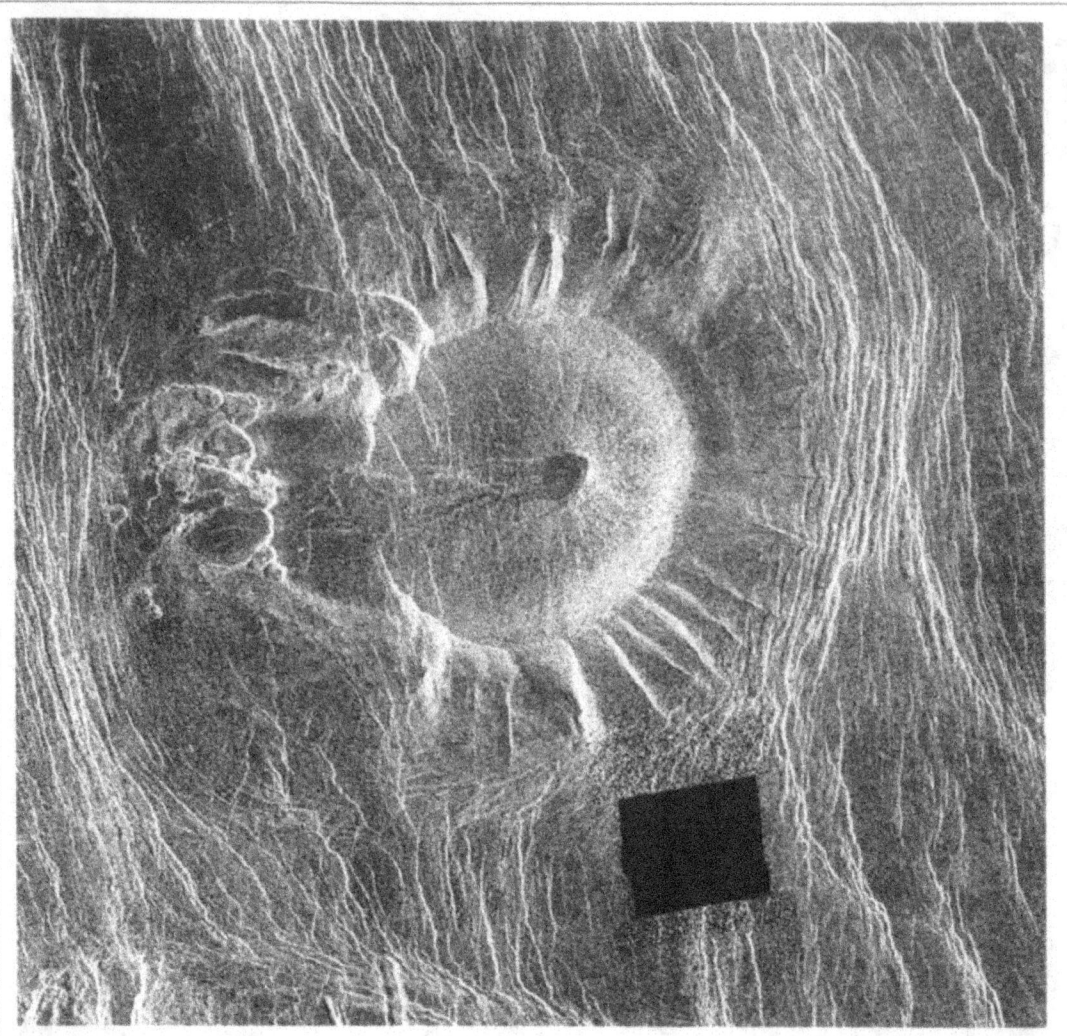

Lying 25 km north of Alpha Regio tessera in the southern hemisphere of Venus, this feature measures 122 km across and 107 km north to south. The summit of the feature is slightly concave and contains a shallow pit near its center. Radar-dark lava flows traveled from the pit westward over the rim, defining this as a volcanic edifice (informally dubbed a "tick" because of its likeness to the arachnoid). Directly west of the rim lies a series of pit craters formed by the collapse of crust over near-surface magma. The scalloped edges seen in this image are a definitive characteristic of ticks on Venus. Although in many cases the edges are thought to have formed by the collapse of material at the edge of the volcanic construct, another possibility is that they are the result of the lateral injection of lava and buildup of ridges. The size, shape, and steepsides of some ticks suggest that they may have once been steep sided domes before experiencing modification. Like the domes, ticks are usually associated with tessera and coronae. (The black area in the image represents missing data.) **5.38**

This scalloped margin dome (or "tick"), located in southeast Beta Regio, is approximately 30 km across and has a flat summit containing a crater 6 km in diameter. This crater, as well as the entire edifice, is surrounded by faint concentric lineaments that suggest subsidence along faults. The edges of this feature have been scalloped and steepened by collapse of the rim material. The material has formed a radar-bright fan deposit west of the edifice, similar to deposits surrounding the large volcano Olympus Mons on Mars. Small volcanic domes are scattered northeast and southeast of the tick. **5.39**

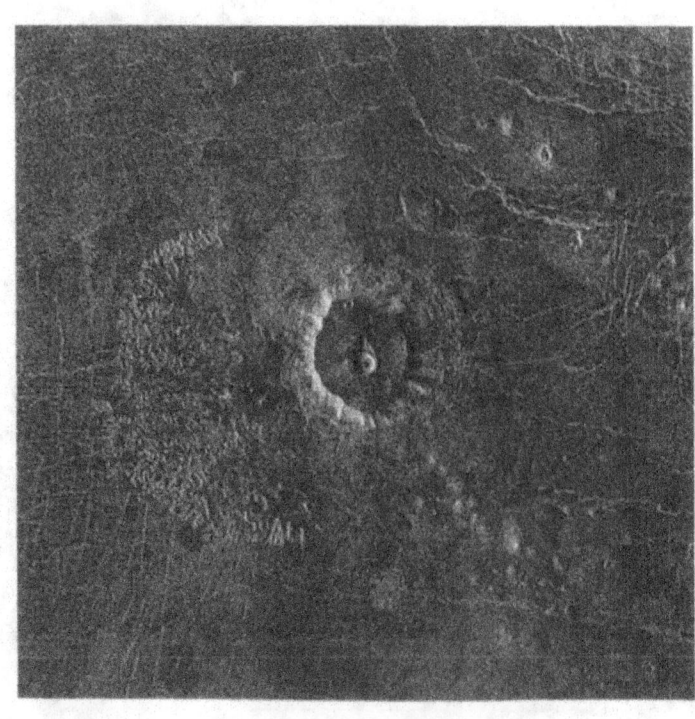

Small volcanoes on Venus number in the hundreds of thousands to millions

Two domes lie in the plains just east of Beta Regio. The southeast dome is approximately 45 km across and is dominated by a caldera 20 km in diameter. The edges of this dome are scalloped, and faint, radar-bright landslide deposits can been seen to the east and southwest of the edifice. This particular "tick" is surrounded by concentric scarps, suggesting that it has undergone subsidence, perhaps due to the withdrawal of magma from beneath the volcanic edifice. Further to the east, bright, fin-ger-like lava flows extend away from the dome. The dome to the northwest of the image has been extensively modified by thermal erosion caused by lava flows producing deep, linear pits that radiate from within the dome. The lava flows that probably caused these channels are difficult to see but are likely related to the variations in brightness of the material surrounding the domes. **5.40**

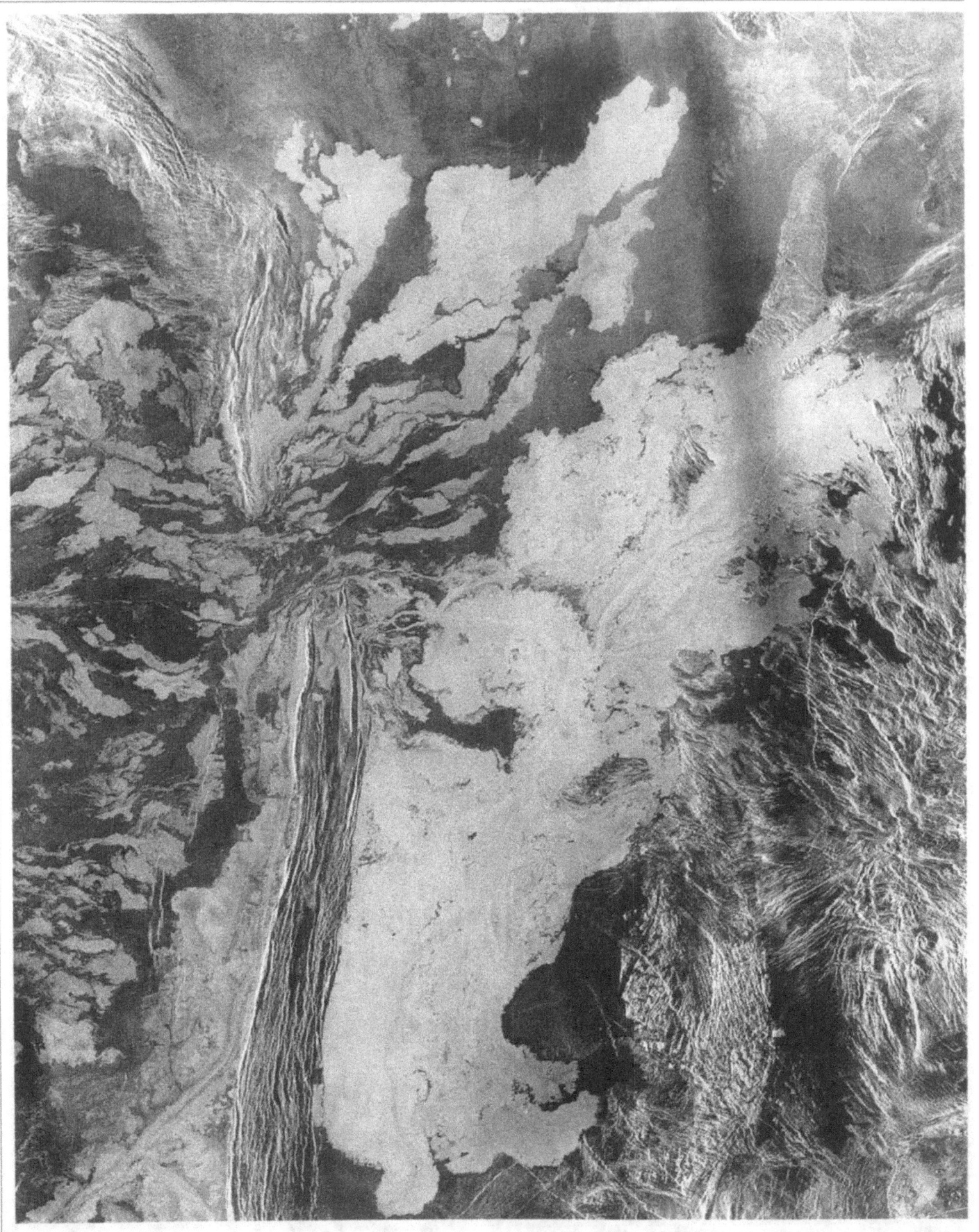

Flows and Channels

With over 80 percent of the surface of Venus covered by volcanic deposits, it is clear that even continuous eruptions from very large volcanoes could not have produced enough lava to cover so much area. The improved image resolution provided by Magellan allows details of Venus' volcanic plains to be observed. The massive outpourings of plains-forming lavas in giant flows and along extensive channels must surely have been an incredible sight.

Located in the Lada Terra region of Venus (–47.0°, 25.0°), this voluminous lava field comprises radar-bright and radar-dark lavas that have flowed over 680 km east from their source area (Ammavaru Caldera, 300 km west of this image). The flows have breached a north-trending ridge belt and ponded to form a vast, radar-bright deposit. The total area for this flow field exceeds 500,000 square km and represents one of the largest or "great" flow fields on Venus. It is similar in scale to many large flow fields on Earth known as flood basalts, such as the Columbia River Basalt Province in central Washington and the Deccan Traps in India. Many of the flows contain central channels similar to (but much larger than) typical basalt flows on Earth. 6.1

5 kilometers

Lava flows on Venus are very similar in gross morphology to lava flows on Earth and the Moon and on other terrestrial planets such as Mars. Unlike flows on Earth, however, lava flows on Venus are typically of a much larger scale. Venusian lava flows quite often reach lengths of several hundred kilometers, and in places exceed 1000 kilometers in total length. Widths generally range from several kilometers to tens of kilometers. On Earth, lava flows rarely exceed 50 kilometers in total length and flow widths typically are measured in meters. Venusian flows are therefore similar in scale to some of the very largest eruptions on Earth, those associated with massive outpourings of lava known as "flood basalts" (which are inferred to produce lava flows hundreds of kilometers in length). On Earth, such eruptions are rare (which is fortunate for civilization!) and are thought to be related to unusual thermal conditions in Earth's mantle. Why lava flows on Venus are frequently so much larger than those on Earth is a phenomenon not yet completely understood. Higher atmospheric temperatures on Venus decrease the

cooling rate of lava flows somewhat, but not enough to account for the differences in flow length observed on the two planets.

Lava flows on Venus are for the most part thought to be basaltic in composition and characterized by morphologies suggesting relatively fluid lavas. Basalt flows on Earth are typified by two dominant surface textures. "Aa" flows are rough with jagged, crinkly surfaces (Fig. 6.2). "Pahoehoe" flows are relatively smooth with "ropy" surfaces (Fig. 6.3). Variations in surface roughness affect the relative brightness of flows in the Magellan radar images (rougher surfaces appear brighter) and may reflect differences between aa and pahoehoe flow types. Such variations also may reflect differences in the age or state of preservation of separate flows. As on Earth, central channels and lava tubes (channels that have cooled and roofed over) are common features of venusian flows.

Flow fields on Venus are associated with a variety of volcanic vents. Central volcanoes are typically surrounded by an apron of flows that constitute the vol-

canic edifice In many cases, lavas erupt in massive sheets that flood the surrounding plains and produce large flow fields that are not necessarily part of a central volcanic edifice. In addition to large volcanoes, flow fields are also associated with fissure vents, coronae, dense clusters of smaller volcanic domes, cones, pits, and channels.

It was known prior to the Magellan mission that the present environment on Venus would not support liquid water on the surface. One of the goals of Magellan, however, was to search for evidence for the presence of water over the history of the planet. This could have taken the form of precipitation, run-off, and the formation of channels in periods of venusian history when the environment might have supported liquid water. It is also known from the planetary record that channels could have formed in association with lava flows (the central lava channels typical of many lava flows on Earth) and with high-effusion-rate, long-duration lava eruptions that thermally erode the substrate and create deep sinuous valleys (such as the rilles of the Moon).

Incoming Magellan images were scrutinized in detail for evidence of channels and channel-like features, their characteristics and ages, information about their source regions, and deposits at the ends of the channels. The search was rewarded with evidence for a wide array of channel types and associated features. Over 200 channels and valley landform complexes were discovered in the initial analyses. These new features were classified as simple channels, complex channels, or compound channels. Simple channels are characterized by a long, single, main channel. The category includes rilles similar to those observed on the Moon, and a new class dubbed "canali," which are extremely long, relatively narrow, individual channels that display very constant widths along their entire reaches. One of the canali has been mapped for a distance of over 7000 kilometers.

Complex channels include anastomosing and braided patterns as well as branching, distributary patterns. These types of channels have been observed in association with the massive outflow deposits

Venusian lava flows quite often reach lengths of several hundred kilometers

A close look at a pahoehoe lava field in the Hawaii Volcanoes National Park. Although the billowy, ropy texture of the pahoehoe lavas often makes them appear dark in radar images, they can sometimes also be relatively bright, making it difficult to distinguish them from aa flows. Aa and pahoehoe (both terms are of Polynesian origin) constitute two varieties of basaltic lava surfaces. [T. G. Farr, Jet Propulsion Laboratory.]

6.3

occurring on the flanks of many impact craters and in the large outflows of lava associated with tremendous fields of lava flows (such as Mylitta Fluctus). Compound channels have both simple and complex segments; the largest of these have an anastomosing pattern and streamlined hills similar to those observed in many of the outflow channels on Mars. Although compound and complex channels superficially resemble patterns created by running water, there is no evidence so far to indicate that liquid water ever occurred on the preserved surface of Venus, which is estimated to average 200–600 million years in age.

Thus, liquid water apparently was not stable for about the last one-half billion years in the atmosphere and on the surface of Venus. But what carved the spectacular channels? Two candidates seem most likely: lava and flowing impact ejecta. Lava flows can

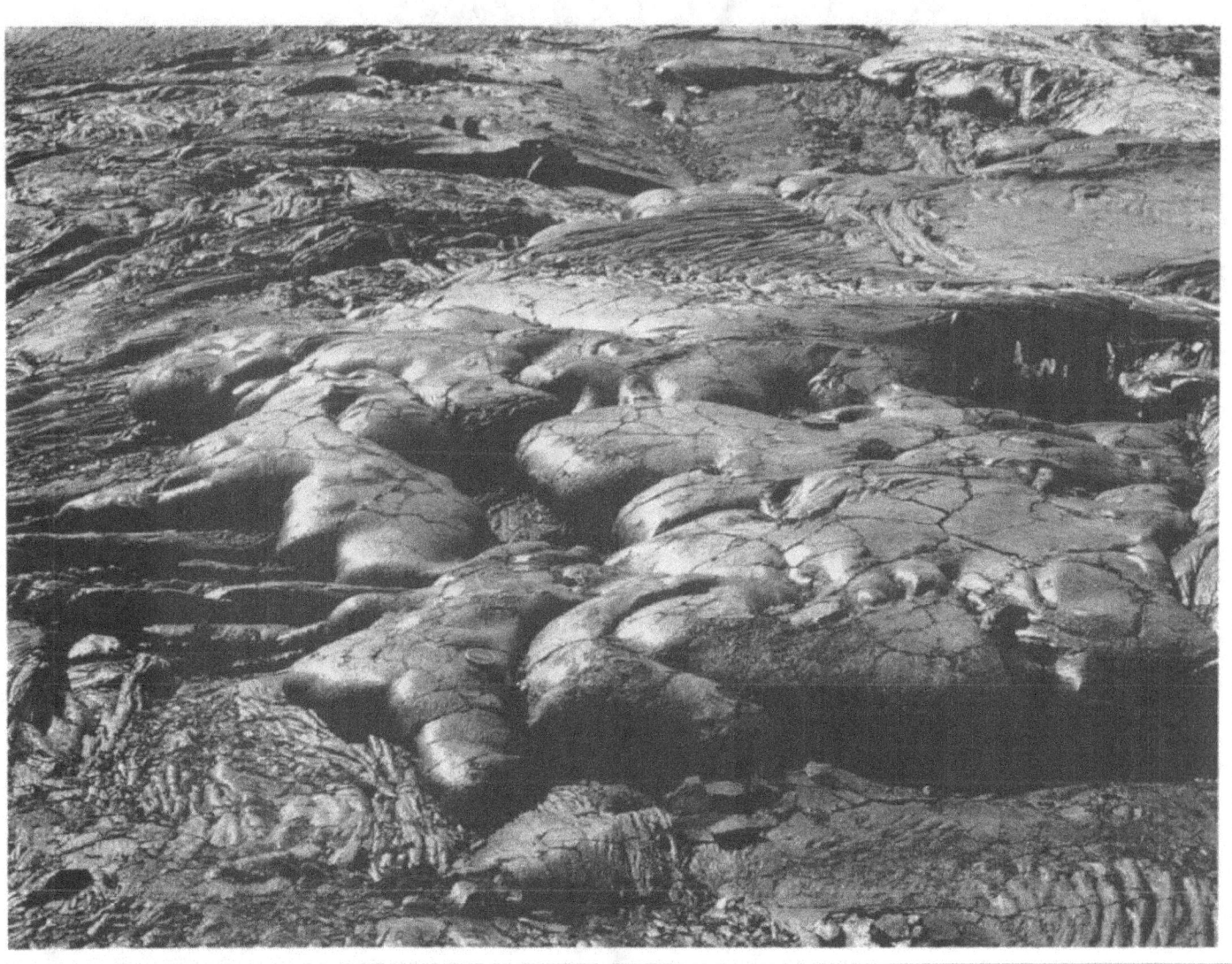

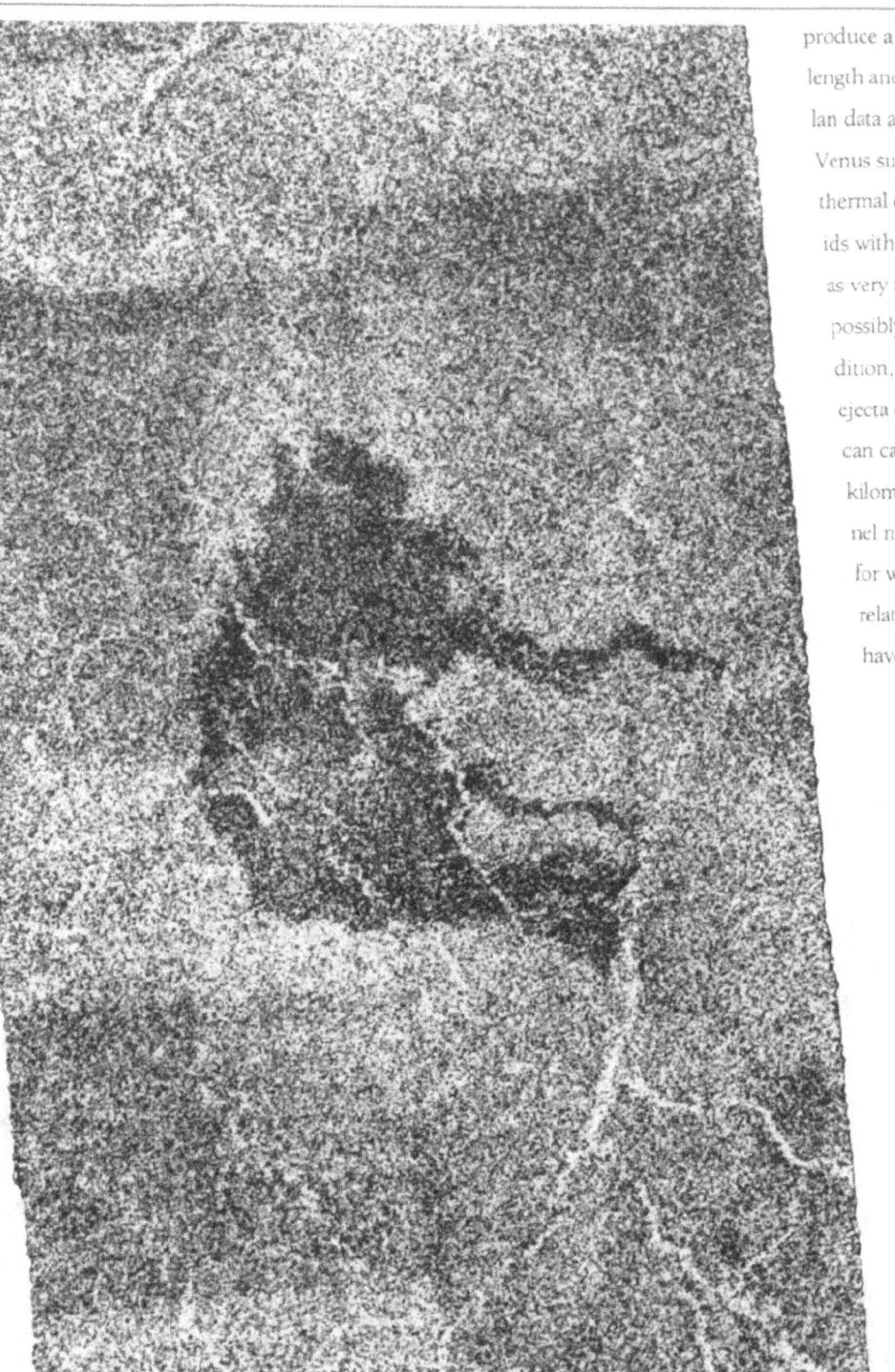

produce a variety of channel forms, but the length and other characteristics seen in Magellan data are highly unusual. Perhaps the hot Venus surface environment helps to encourage thermal erosion. Possibly some exotic lava fluids with very low viscosities are involved, such as very iron- and magnesium-rich basalts or possibly even sulfur or carbonate lavas. In addition, it is clear that the interaction of impact ejecta curtains with the Venus atmosphere can cause flows to extend for hundreds of kilometers and have typical flow-like channel morphologies. Although no evidence for water has been found, new mysteries related to the unique Venus environment have certainly emerged.

These radar-dark lava flows appear to have emanated from a fissure seen in the lower right portion of the image. They have flowed approximately 23 km west–northwest and coalesced to form a larger deposit on older, brighter plains in southern Guinevere Planitia (4.2°, 331.8°). This is a relatively small flow deposit compared to other flow fields on Venus, covering only about 260 square km. 6.4

3 km

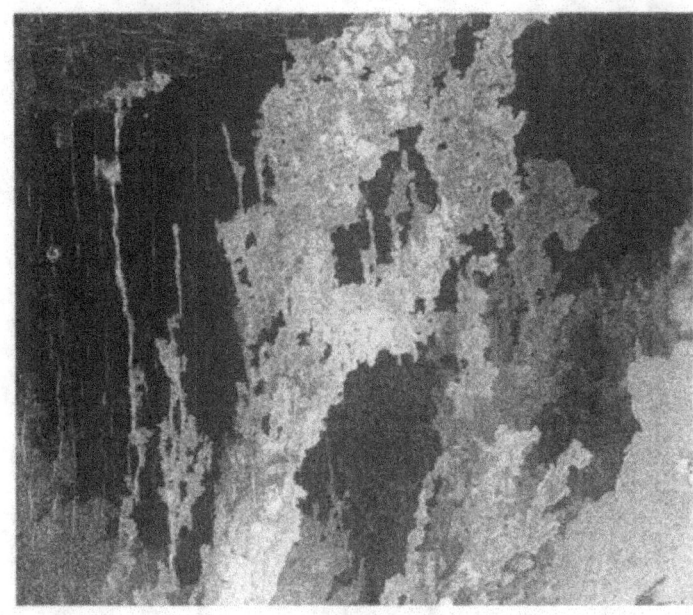

R adar-bright flows from the large volcano Sif Mons in western Eistla Regio, just south of this scene, are shown in this image centered at (25.0°, 352.0°). Some of the lavas forming this flow field were directed down preexisting narrow fissures or graben. In contrast to Fig. 6.4, the lavas merely flowed down — and were not fed — by the graben. The flows within the narrow fissures are on the order of 1 km in width, compared to the larger flows (not contained within the fissures) that are 10–20 km in width. Flow lengths shown in this image are approximately 140 km; total flow lengths from the volcano summit to the outer flanks exceed 400 km. The great lengths and confinement of flows to narrow graben suggest they were formed by extremely fluid lavas, in contrast to the much more viscous flows of Fig. 6.4. **6.5**

T his highly unusual volcano is located on ridged plains between Artemis Chasma and Imdr Regio (−37.5°, 164.5°). It is characterized by a complex central dome 100 km across from which broad channels 10–15 km wide radiate and terminate in extensive flow lobes 40–50 km across. Additional flows extend up to 400 km to the east from the central structure. The flows are distinctive on the basis of their extremely rough surfaces (composed of ridges spaced 700–800 m apart) and great thicknesses that range up to 700 m. Lava scarps on the east side of the flow field are about 90 m high. These flows appear to be the result of lavas that are much more silicic and/ or viscous in nature, in contrast to the relatively thin, fluid flows of Sif Mons and Ammavaru, which are thought to be basaltic in composition (Figs. 6.1 and 6.5). In addition to the "pancake domes" (Chapter 5), these flows suggest that a wide range of lava types may exist on Venus. **6.6**

40 km

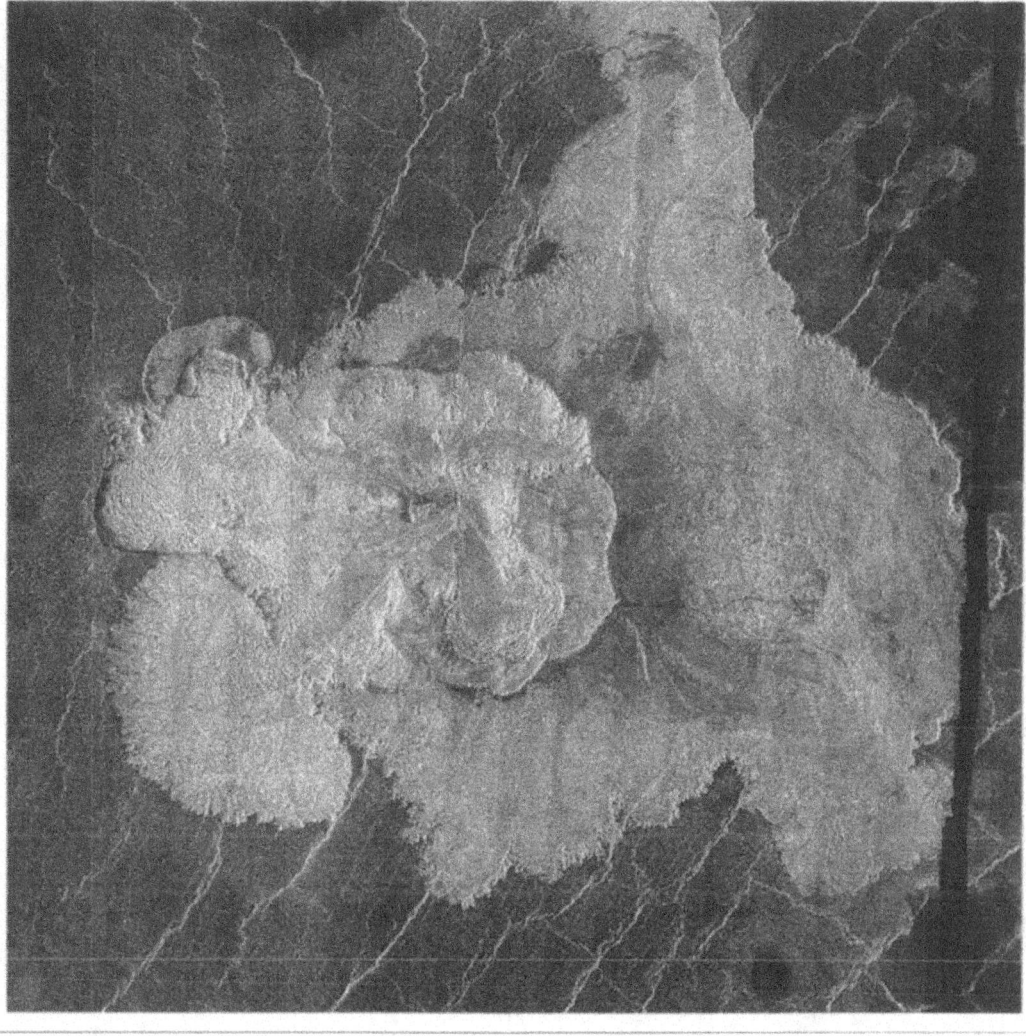

T he fan-shaped flow field of Mylitta Fluctus is shown in this image, centered at (–55.0°, 355.0°) in Lavinia Planitia. The flow field of Mylitta was one of the first large flow fields observed on Venus. The flows originate from a shield volcano situated within a rift belt just south of the image. The northern flank of the shield is visible in the lower left portion of the image. The bulk of flows composing the flow field are superimposed on the shield and have flowed down-slope north into the plains of Lavinia to form a flow field that is approximately 400 km by 1000 km in total extent. The main flow field descends 1 km from the shield into the plains where it encounters more level ground and the flows pond and widen. Individual flows range in width from several kilometers to over 120 km. A variety of flow features such as braided channels, levees, and changes in flow-surface texture can be seen in the lower portion of the image. **6.7**

——————— 75 km

S harp contrast in flow morphology is seen in this image of flows along the southeast flank of Ozza Mons in Atla Regio. In the upper part of the image are flows that are only a few kilometers in width and up to 200–300 km in total length. Such flows are typical of many volcanic shields on Venus. In the central portion of the image are flows of remarkable size, up to 75 km in width and extending over 1400 km in total length. Within these larger flows are distinctive channels with well-developed levees. Some regions of the flow surface appear to have been broken up, forming the unusual bright and dark patterns adjacent to portions of the central channel. Differences in flow morphology as expressed in these two flow fields are probably the result of differences in the total volume and rate at which lava is erupted from a volcanic vent. Very large volumes of lava erupted at a high effusion rate are able to flow for great lengths before solidifying and producing flows of the scale observed in the central part of this image. **6.8**

——————— 100 km

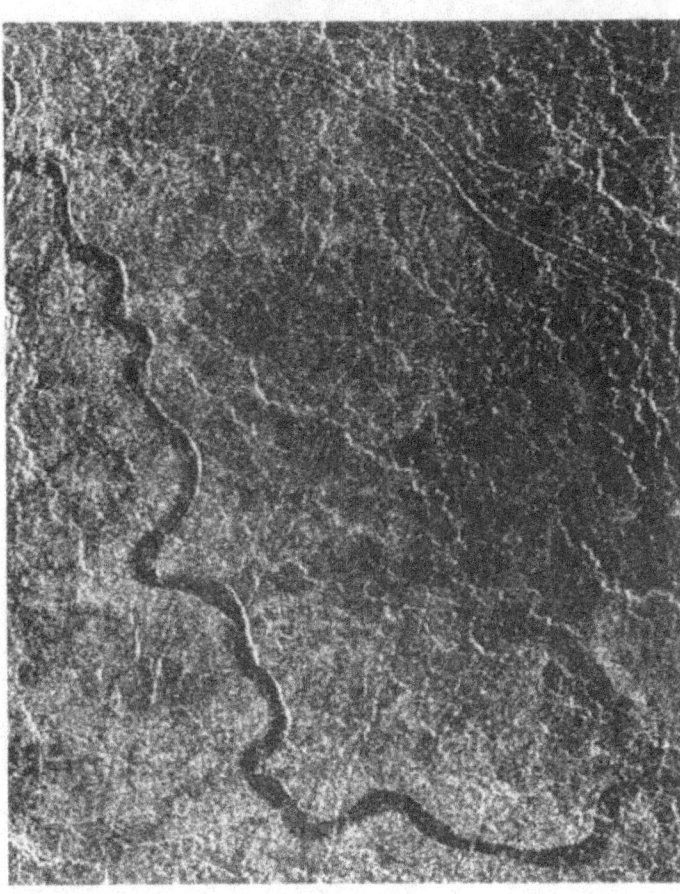

This segment of a simple canali-type channel is only a few kilometers wide and is located in Helen Planitia (centered near −49.0°, 271.0°). It is part of a long channel that extends for hundreds of kilometers. The plains in the image are formed of lava flows and the dark channel appears to have been incised into the background plains, suggesting the kind of thermal erosion that occurs in lunar rilles. Irregular bright lines oriented northwest–southeast and north–northeast are fractures and faults. Some of these cut the dark floor of the channel, suggesting that deformation postdates the channel formation. A faint channel of similar width, but not filled with radar-dark areas, is seen in the upper right part of the image. 6.9

20 km

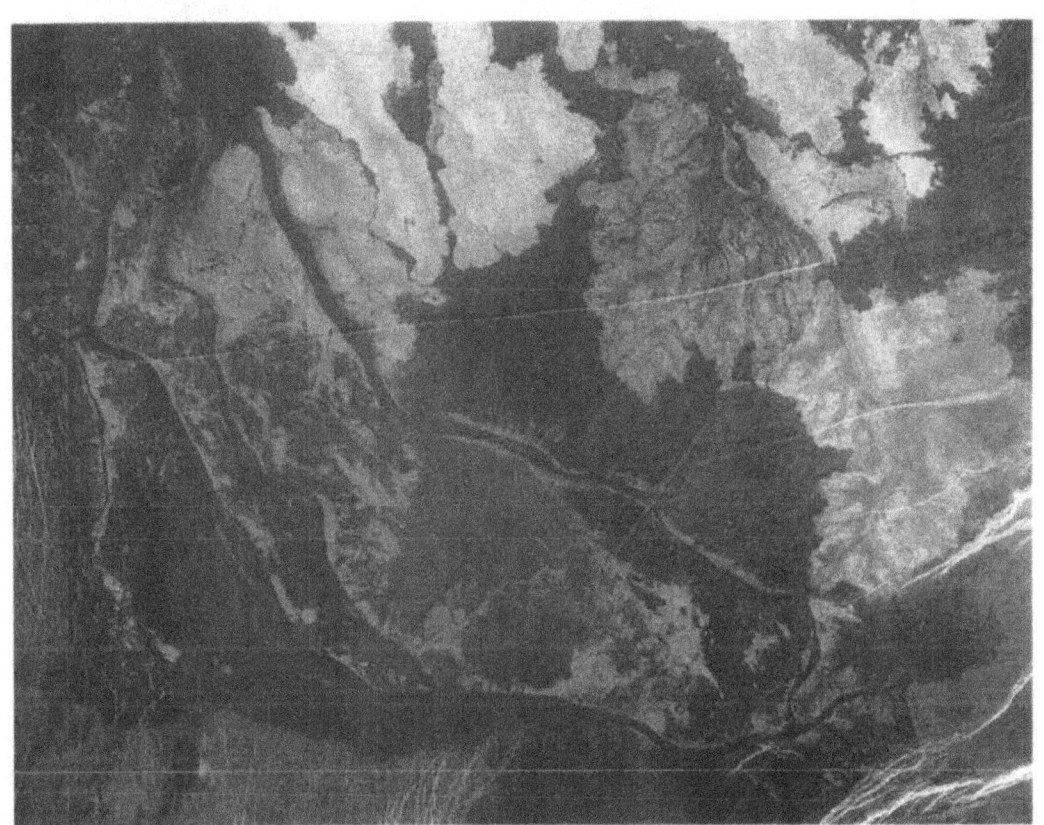

A flow field with a complex network of branching channels is shown in this image of a region southeast of Aphrodite Terra (−20.0°, 180.0°). Lava flows in this area are dominated by central channels that are typically radar-dark and branch downstream (towards the bottom of the image) to form delta-like patterns. Individual channels branch and narrow downstream, which is in contrast to channels cut by running water — such as rivers and streams on Earth — and more typical of channels formed in lava flows. 6.10

75 km

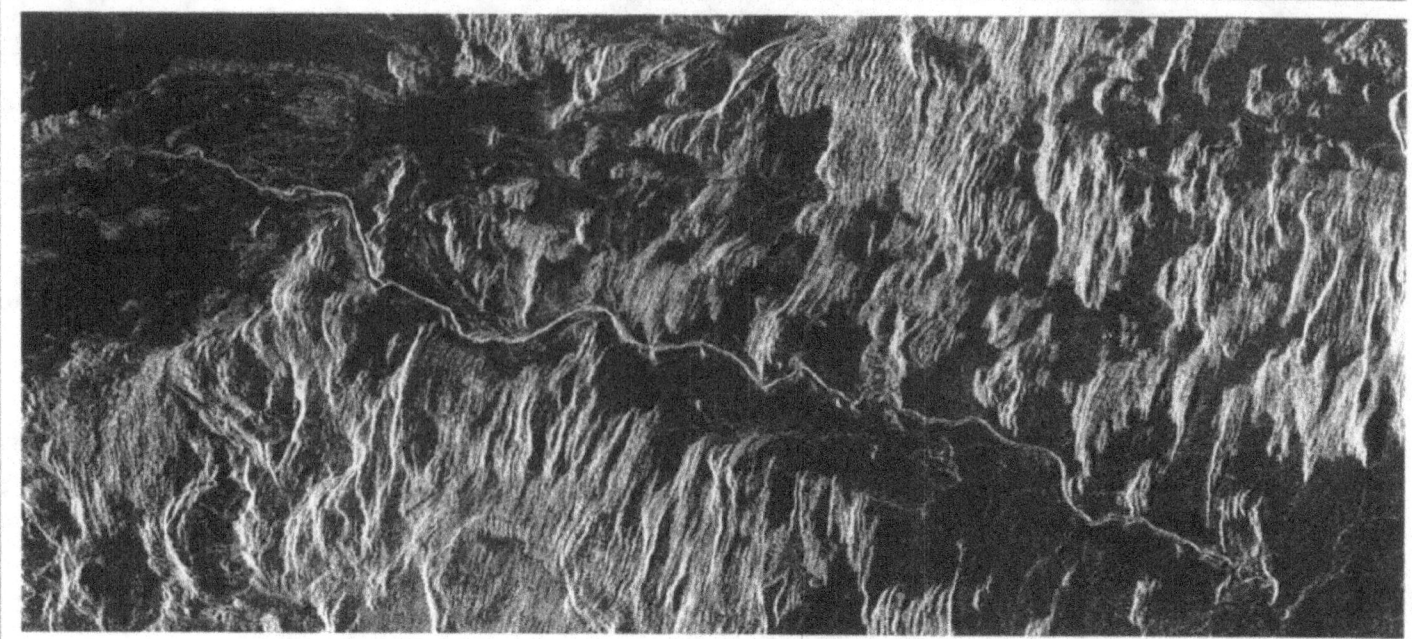

Venus' present environment cannot support liquid water on the planet's surface

A sinuous channel about 2 km wide emerges from a circular depression in lava plains within Atropos Tessera at the western end of Ishtar Terra (68.5°, 309°) and flows westward through the complex tessera terrain down toward the surrounding plains for a distance of over 180 km. The channel is clearly of volcanic origin and empties out into a series of flow-like features seen in the upper left part of the image. 6.11

▬▬▬▭▭▭▭ 15 km

Integrated valley network located at about (2°, 70°), north of Ovda Regio. Such integrated networks form a complex system of branching valleys that have morphological similarities to features formed on Earth and Mars by the process of sapping, causing the removal and undermining of material and its subsequent collapse. This feature is over 50 km in length and the individual channels are less than 1 km wide, forming a series of apparent tributaries. The network occurs in the midst of a patch of radar-dark lava plains. 6.12

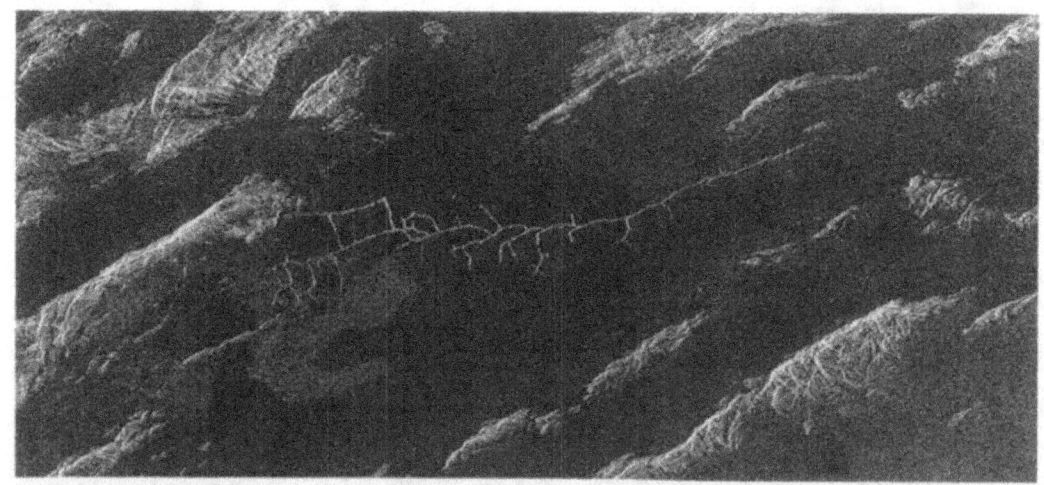

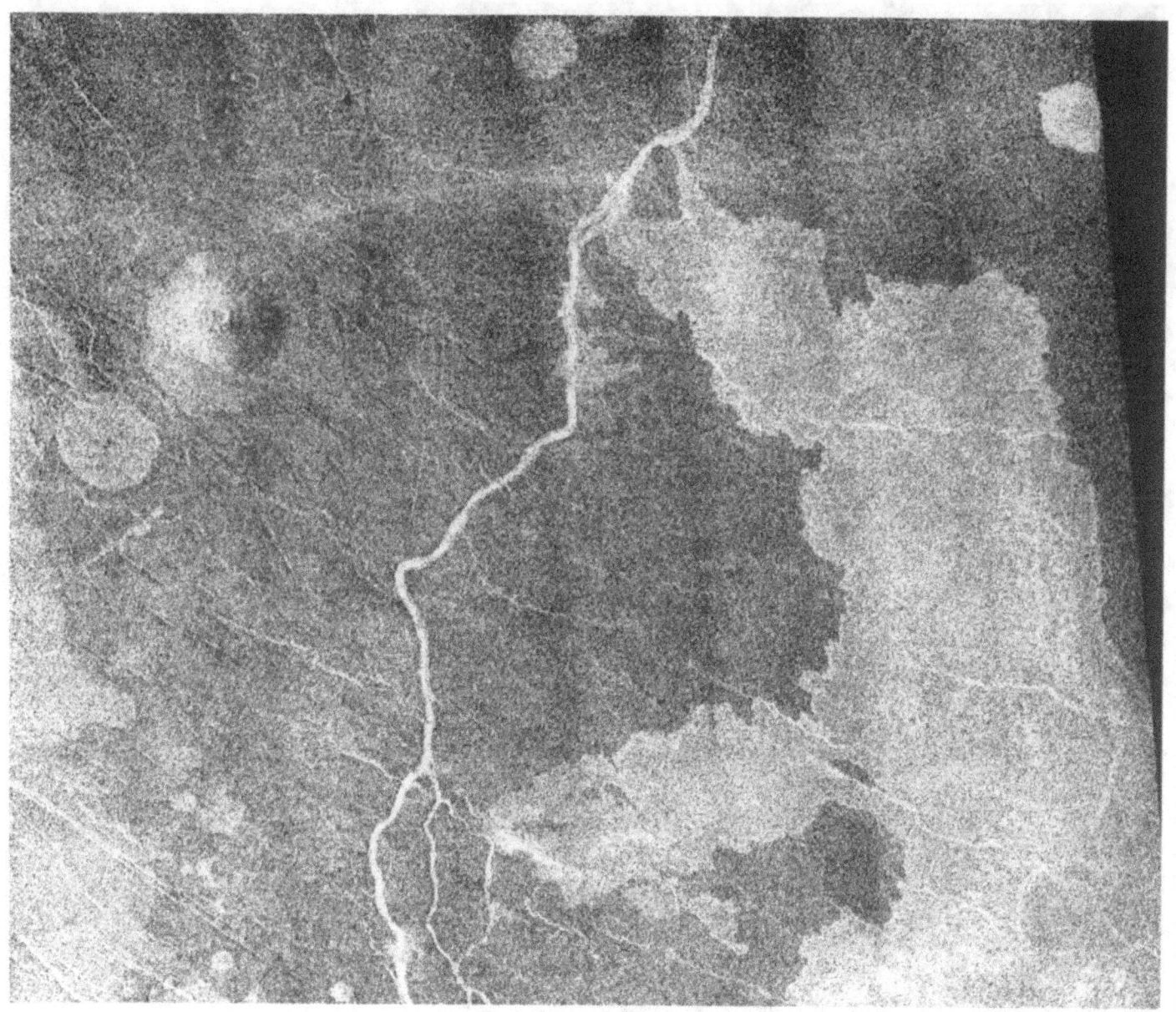

In some cases, material apparently spilled out of sinuous channels as it was being transported great distances across the venusian plains. In this example, a radar-bright sinuous channel less than about 2 km across winds its way across the volcanic plains of southern Guinevere Planitia (6°, 319°). On the left are several small shield volcanoes less than about 20 km in diameter. In the north–central part of the image, radar-bright lava flows emerge from two points along the sinuous channel and extend out into the surrounding plains, flooding a large area in the eastern part of the image. This spillover provides evidence that the channels were transporting lavas with different radar properties than the surrounding plains for significant distances across the surface of Venus. 6.13

━━━━━━━━━ 15 km

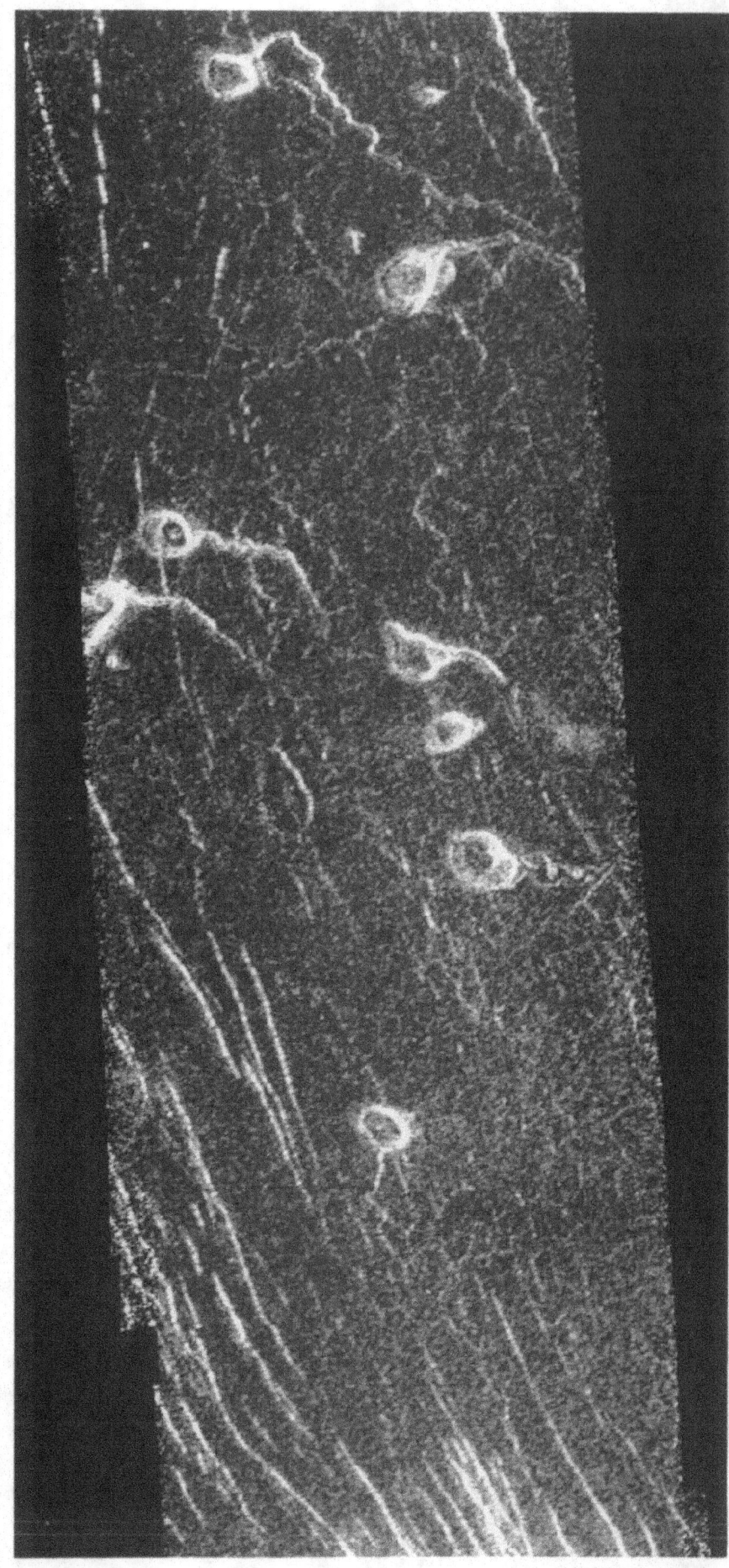

Source regions of the channels often provide evidence for a volcanic origin of the channels themselves. In this part of Lavinia Planitia (–29.6°, 341°), a series of tadpole-like features is interpreted to represent lava sources (the circular depressions) where lava has emerged to the surface and flowed generally eastward down the regional slope, forming a series of lava channels that extends down into the surrounding plains for distances in excess of 10 km. A variety of source craters up to nearly 3 km in diameter can be seen, ranging from circular to irregular in shape and with steep sides and flat, lava-flooded floors. **6.14**

5 km

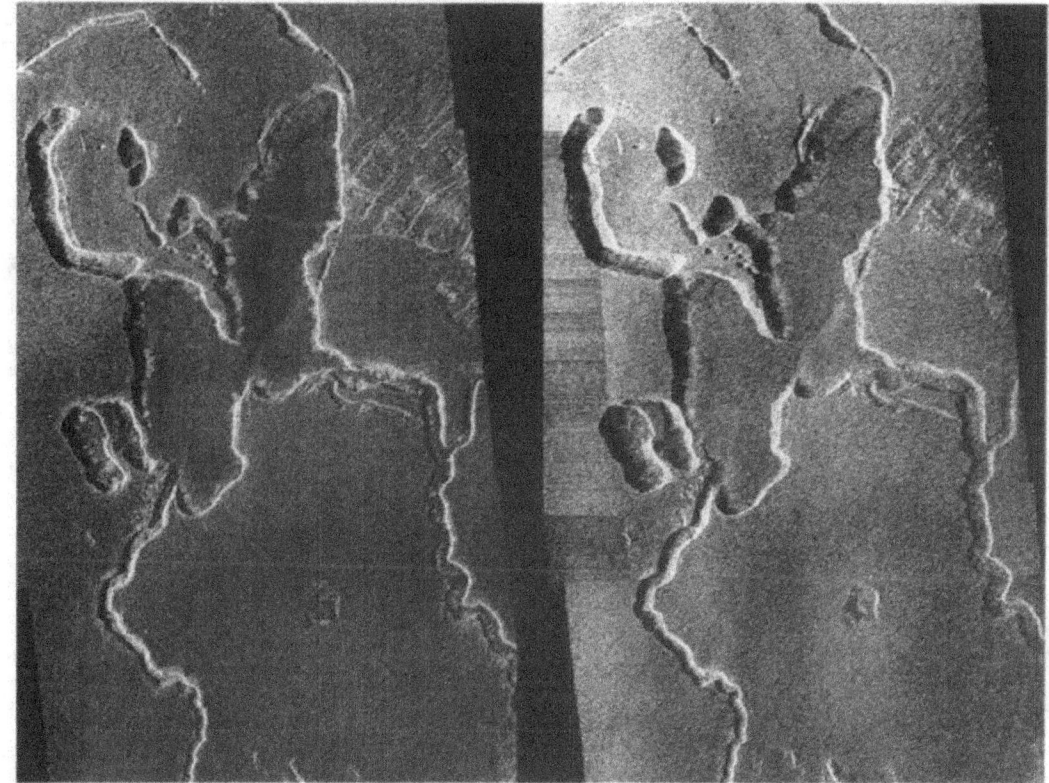

In this stereo pair of a large irregular depression and associated channels located in southern Ovda Regio (–11.6°, 88.7°), the relationship of source regions and channels is further clarified. The irregular, lung-shaped depressions are source regions and vents for the lavas and were probably lava ponds at the time of the eruptions that created these features. Lava left the ponds along breaches at the lower left and right and carved extensive channels extending for hundreds of kilometers into the surrounding terrain. As the eruption waned, the lava pond eventually drained through these channels, leaving the presently observed topography. 6.15

A segment of the longest channel yet found in our solar system winds its way through southern Atalanta Planitia (49°, 165°). This channel (highlighted by arrows), about 1.8 km wide and more than 7000 km long, is several hundred kilometers longer than the Nile, the longest river on Earth. The length of the portion seen in this image is about 600 km. Although neither end of the channel is exposed, the morphology is similar to others on Venus that appear to have a lava-channel origin. The extreme length, narrowness, and consistent width are not yet accounted for. How can lava flow over such great distances in such narrow channels without cooling and solidifying? Even without knowing the detailed origin of the channel, it was apparently carved by fluids. This knowledge provides us with a measure of the amount of deformation of the surface on which it was emplaced after it had formed. For example, if the channel was originally emplaced on a flat surface (as suggested by its meandering surface), several places where it appears to go uphill and downhill (for example, in the generally north–south oriented ridge belt and in the circular feature along the western margin of the image) probably postdate the emplacement of the channel. The network of fine-scale deformation ridges in the plains also appears to postdate the formation of the channel because the channel does not seem to be influenced by their topography; several ridges are seen to cut across the channel. The image is 460 km in width. 6.16

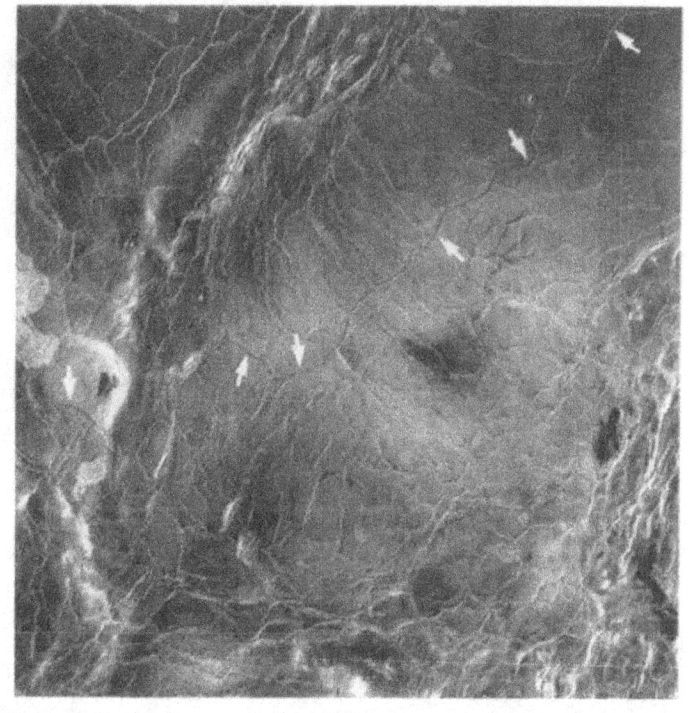

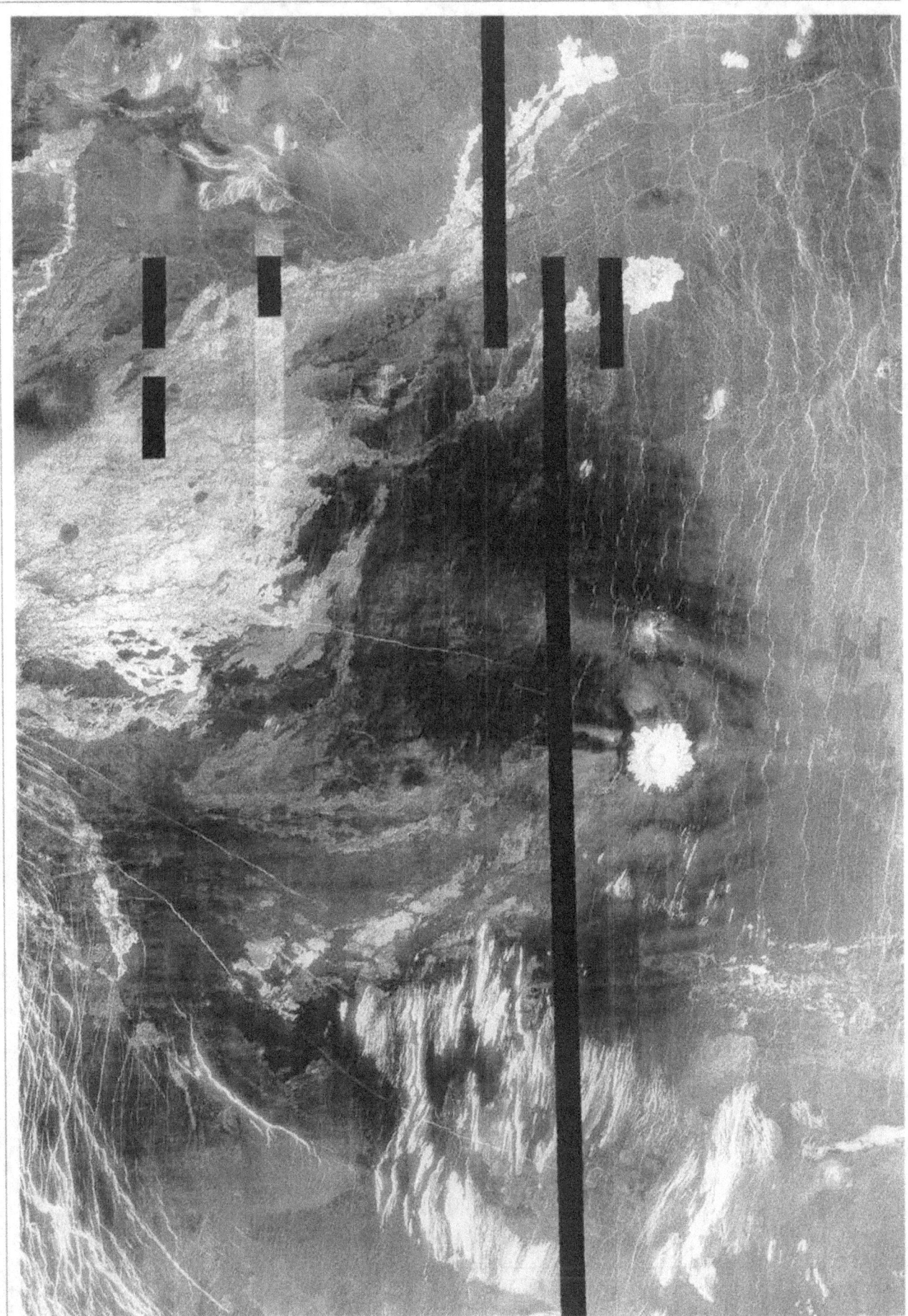

Surface Processes

Lack of water on the planet's surface has made erosion and deposition much less effective geological forces on Venus than on Earth. Over the last several hundred million years, wind has generated countless streaks, dunes, and wind-sculpted hills known as yardangs that cover Venus' surface, while landslides and other mass movements have altered Venus' hilly terrains. Atmosphere—surface chemical reactions have modified the outer several meters of Venus' surface. The planet's highlands are extremely reflective to radar, possibly from weathering or the presence of ferroelectric minerals.

On Earth, the hydrological cycle dominates surface processes, transferring water among ocean, atmosphere, and land reservoirs. But on Venus, with its high temperature (approximately 740 kelvins) and atmospheric carbon dioxide pressure, water cannot exist in the liquid phase. In fact, examination of Magellan's radar images has shown no evidence of landforms generated by water. Dendritic channel systems, terraces, alluvial fans, and shoreline features were all searched for, but not found. Magellan data have instead provided with Venus the opportunity to observe a planetary surface on which processes have been dominated by a dense, hot atmosphere without a significant fluid phase.

A clear signature of atmosphere—surface interactions is the presence of impact crater ejecta, which are strewn across much of Venus' surface. Ejecta excavated during impact are dispersed into Venus' upper atmosphere, where strong winds carry the material westward. As ejecta material settles out on to the surface, deposits form in parabolic patterns. These deposits can be superimposed on various landforms and geological units, including lava flows. Examination of Magellan data shows approximately 60 parabolic deposits on Venus' surface that are associated with impact craters. In every case, the deposits cover surrounding materials, suggesting that these parabolic units are the youngest (geological) deposits on the planet.

Ejecta are also a source of material for reworking by Venus' surface winds, whose velocities — inferred from Venera lander measurements — are approximately one meter per second. Given the dense nature of Venus' lower atmosphere, these velocities are more than strong enough to erode and transport the fine-grained, loose ejecta material. Wind streaks, dunes, and yardangs can all be found in abundance in regions covered by ejecta deposits. Streaks form when winds blow across ejecta or volcanic ash draped over topographic obstacles such as volcanic domes. The domes' leeward sides are scoured of loose debris, exposing underlying material with different roughness characteristics (and thus different radar signatures) as compared to the sediments. Dunes form when sand-

Parabolic ejecta deposits from crater Annia Faustina (diameter 22.5 km; 22.1°, 4.7°). Ejecta thrown high into the atmosphere during the impact have been dispersed to the west of the crater by strong easterly winds. The deposit is draped over lava flows from the volcano Gula Mons, which is off the image to the southwest. The parabola extends about 630 km north to south. Fig. 4.14 is another example of parabolic deposits. 7.1

75 km

size sediment is dispersed by winds into wave-like patterns. Yardangs form when wind-driven material carves into friable deposits and generates deep furrows.

Many of the crater-related dune features on Venus may have been formed by local winds generated by the final stages of the impact process. On the other hand, wind features also have been observed far from deposits of crater ejecta. When features clearly associated with impact processes are removed, evidence points to a trend of wind directions toward the equator — a trend suggestive of a "Hadley-cell" circulation pattern within the equatorial to middle latitudes. In all, Magellan radar data confirm the presence of strong easterly winds high above Venus' surface and meridional winds at the surface.

In situ weathering of Venus' surface has also occurred over the several hundred million year (average) age of the surface. For instance, overlapping lava flows can be observed in many places on the venusian plains. (The basic rules of stratigraphy, of course, state that older units are covered by younger units.) Invariably, older flows have a less distinct backscatter contrast than the surrounding plains. This is the case even though each flow has the same

D unes and streaks from crater Aglaonice. Crater Aglaonice is one component of the "crater farm" (Fig. 4.36) identified in Arecibo radar images. 7.2

15 km

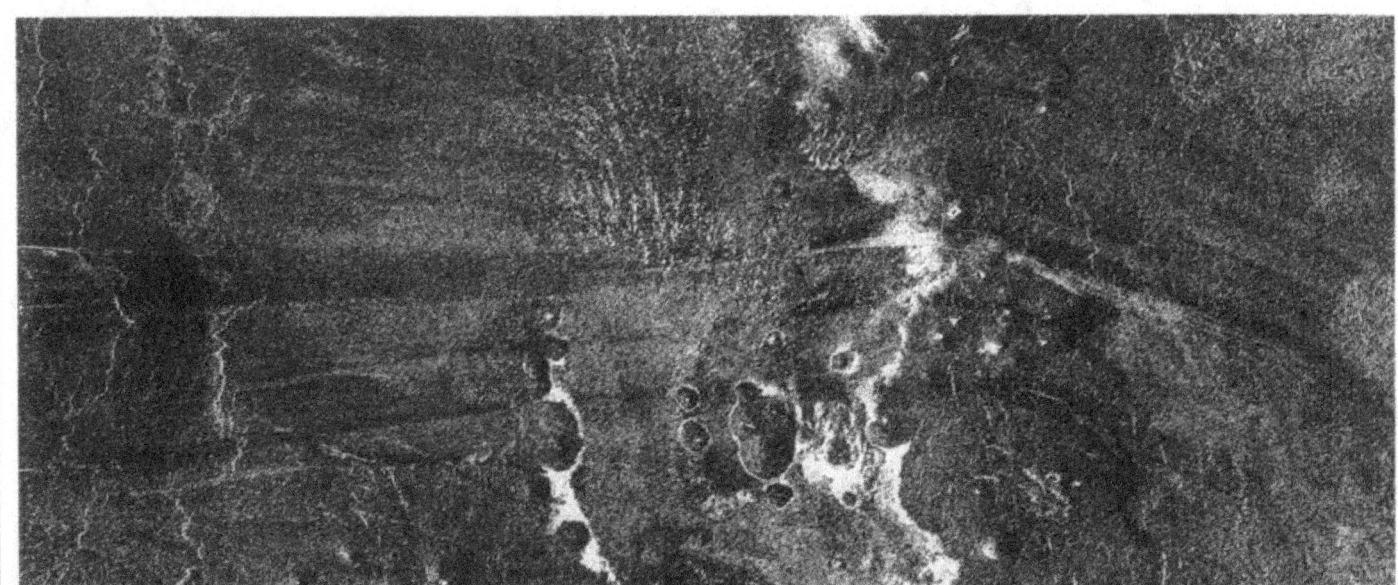

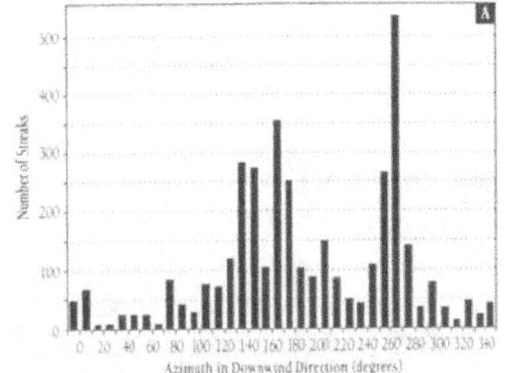

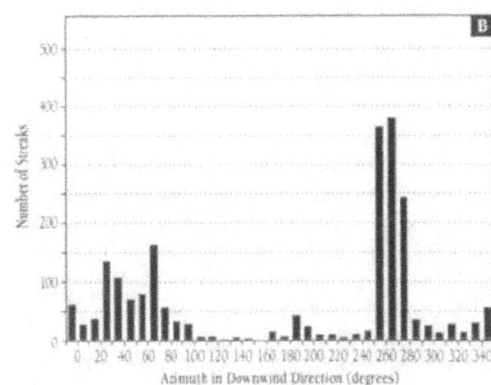

H istograms of the number of wind streaks as a function of azimuth in the northern (A) and southern (B) hemispheres of Venus. North is at 0°. The wind-streak distributions are bimodal, with the dominant wind direction toward the equator and the west in both hemispheres. Figure 7.5 is a schematic representation of the corresponding atmospheric circulation pattern. [R. Greeley, Arizona State University.] 7.3

Crater Mead has associated yardangs to the south and other linear streaks to the north and west. This view is to the southeast of Mead and shows yardangs. Fig. 4.24 presents a full view of Mead. 7.4

20 km

shape. In fact, comparisons of Magellan radar data observations with terrestrial lava flows indicate that the youngest venusian lava flows have signatures comparable to "aa" and "pahoehoe" flows on Earth (Figs 6.2 and 6.3). Older flows, meanwhile, closely resemble terrestrial flows in arid regions smoothed by weathering.

What has caused the chemical and mechanical smoothening of older lava flows on Venus? Atmosphere–surface reactions involving carbon dioxide and sulfur dioxide may be the causes: these two gases are the planet's most-abundant and third-most-abundant atmospheric species, respectively. (Inert nitrogen is second-most abundant.) Possible reactions include the weathering of silicates by carbon dioxide to produce carbonates and quartz, and by

sulfur dioxide to produce anhydrite (calcium sulfate) and carbon dioxide. Until landers are placed on Venus' surface and measure its actual mineralogy, specific descriptions of these reactions are speculative at best. Nevertheless, it is interesting to note that these reactions can weather Venus' surface and buffer and control the planet's overall atmospheric composition.

In terms of processes, one of the visually striking features is the increase in radar backscatter in the venusian highlands to extraordinarily high values above approximately 6054 kilometers in radius. As noted in Chapter 3, this change correlates with a decrease in emissivity, which is a measure of the surface's electrical properties through the dielectric constant. This pattern must be related to the major decrease in temperature with increasing elevation. Several hypotheses exist to describe the surface properties needed to explain such extraordinary radar signatures. One idea is that Venus' surface consists of loose soil with hollow spherical voids that produce very efficient radar backscatter. Another is that the surface is rough and coated with a material that has an extremely high dielectric constant. Yet another theory suggests that the outer meter or so of the surface contains flakes of a conductive mineral such as pyrite. Finally, a recent model that makes very few assumptions holds that Venus' surface contains everywhere a small proportion of what is known as a "ferroelectric" mineral.

Ferroelectric minerals display a unique property: at a critical temperature, the dielectric constant exhibits an abrupt increase. As the temperature increases further, the dielectric constant moves back to normal values. Minerals such as perovskite and pyrochlores, which are ferroelectric in nature, both have critical temperatures of the magnitude required to explain Venus' low-emissivity highlands. The presence of these minerals on the highlands would replicate the transition to low emissivity with increasing elevation (i.e., cooler temperatures) observed on Venus; it would also explain the abrupt transition to normal emissivity values at the highest elevations. Until in situ observations of mineralogy and chemistry are available, however, the

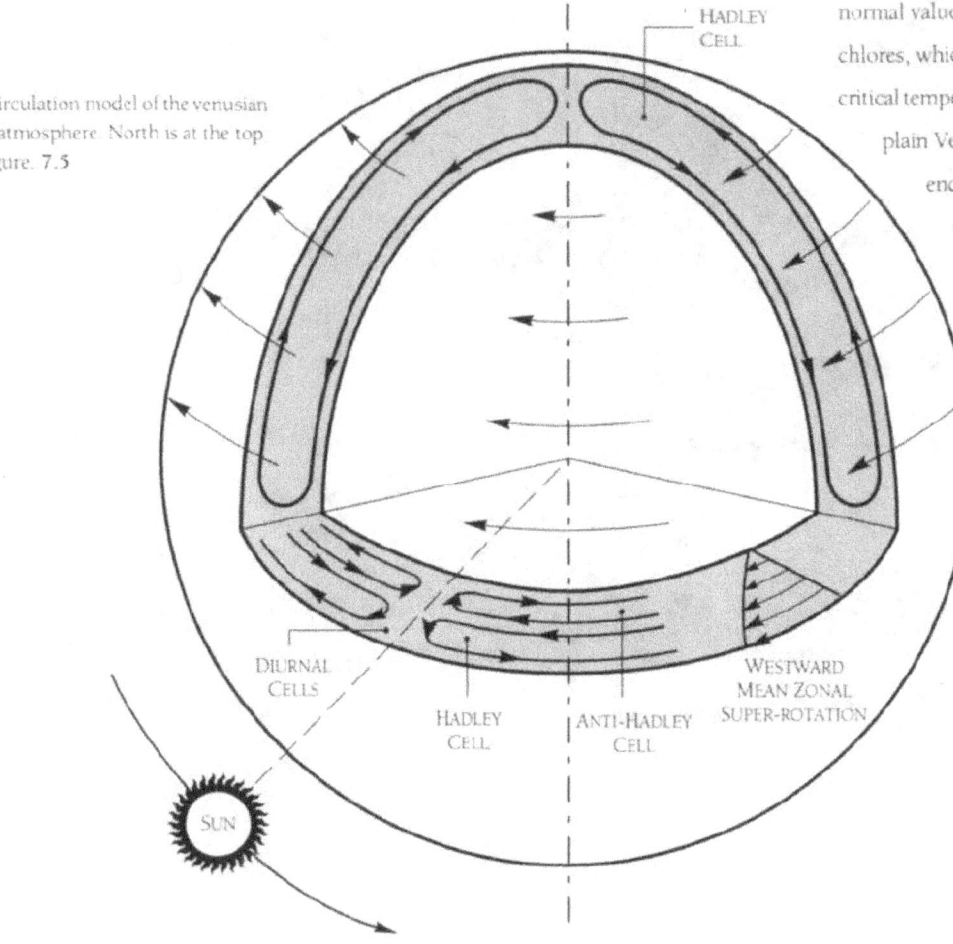

Circulation model of the venusian atmosphere. North is at the top the figure. 7.5

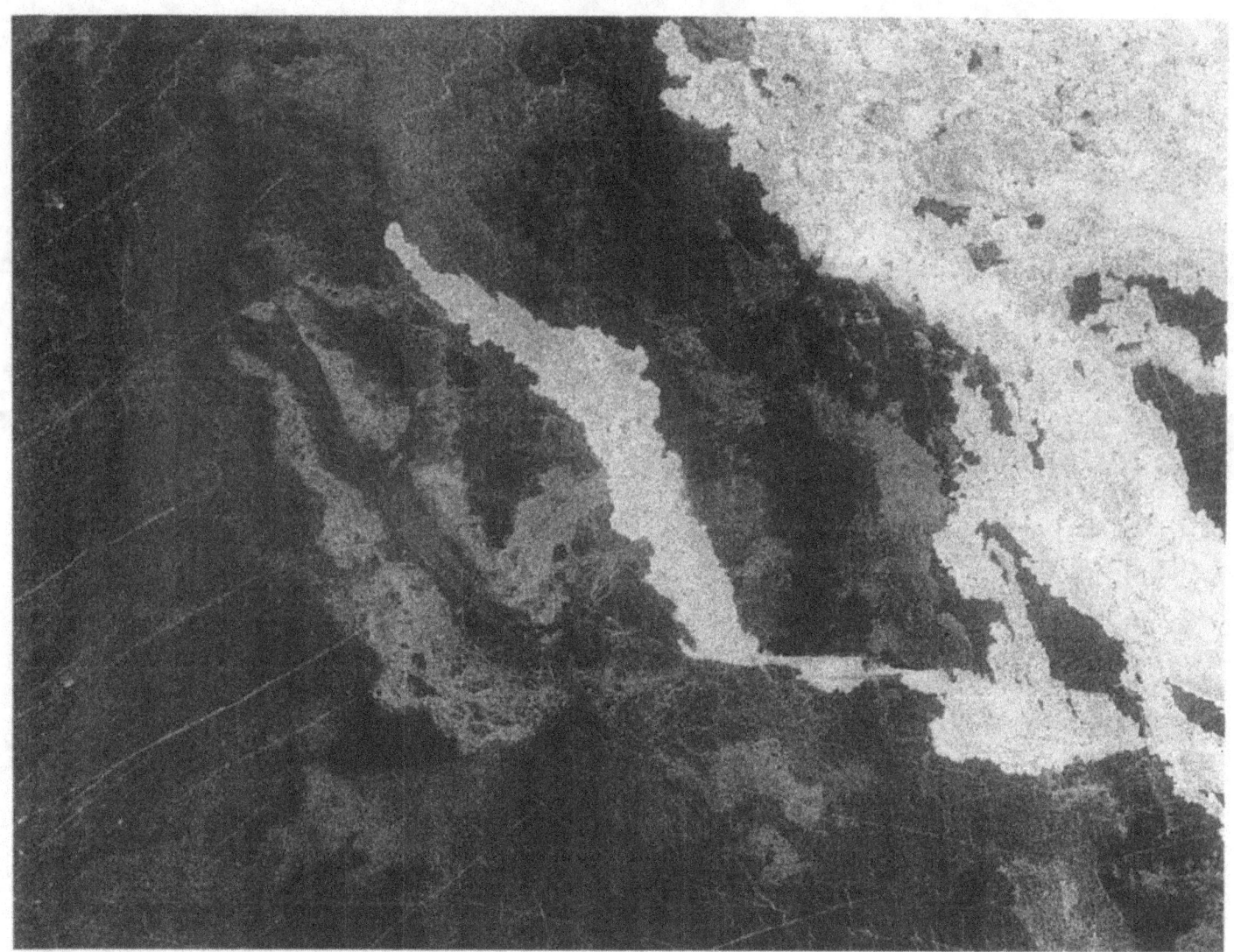

presence of ferroelectric minerals on Venus' surface must remain a hypothesis.

In summary, the lack of water on Venus offers an opportunity to explore the effects — over geological time — of global-scale reactions between the planet's atmosphere and surface materials. Magellan observations offer tantalizing clues to the types of reactions that may pertain. Only further, detailed exploration will allow us to understand the nature and rates of Venus' atmosphere–surface interactions.

Degraded lava flows from Maat Mons. Bright flows are seen over older flows, which in turn cover the low backscatter (dark) plains. The degradation of lava flows is probably the result of in situ weathering from atmosphere–surface chemical interactions as well as erosion and deposition by winds. Fig. 5.1 features a three-dimensional reconstruction of Maat Mons. **7.6**

20 km

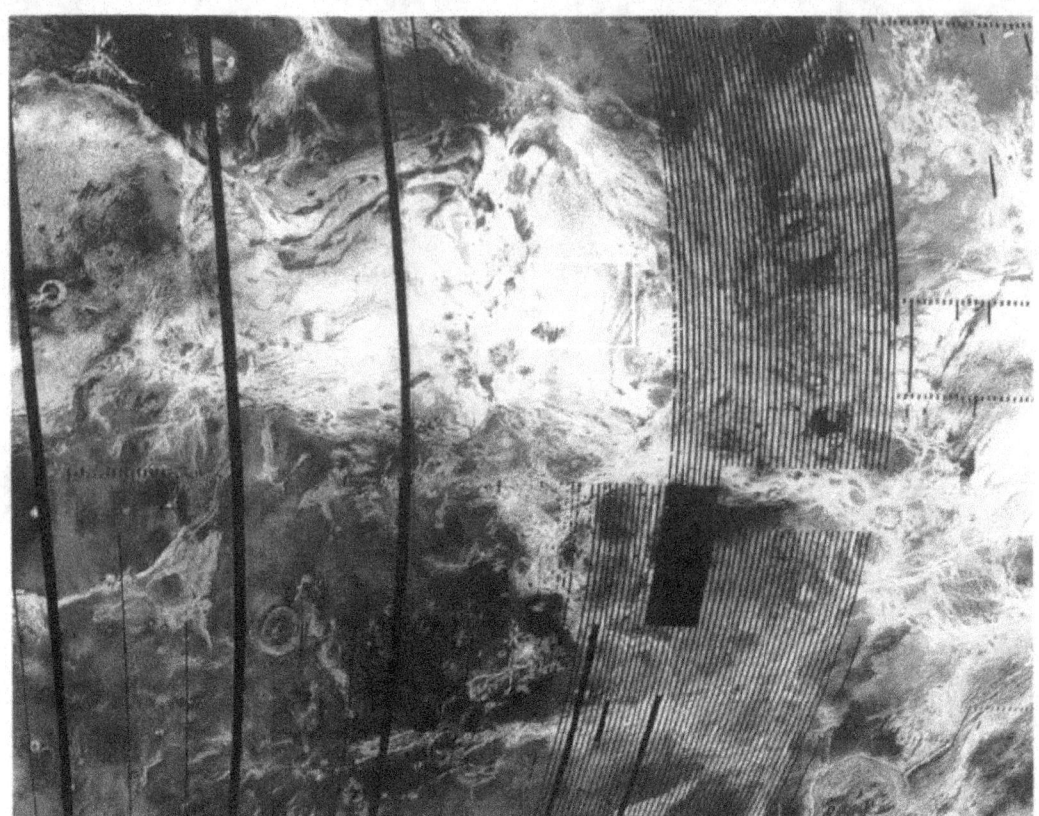

O verall view (top) of the tessera (see Chapter 8) in Ovda Regio. Much of the Ovda highlands lie above the 6054-km radius. The anomalously high radar backscatter makes the region appear bright. The 6054-km cutoff is not sharp. Radar characteristics change over 1–2 km of altitude. The outlined area is enlarged in Fig. 7.8. **7.7**

500 km

C lose-up view of Fig. 7.7 shows an extensive, thick lava flow at the lower left. The dark feature at the bottom left is in fact higher than the lava flow. Fig. 7.9 is a close-up of the outlined area. **7.8**

75 km

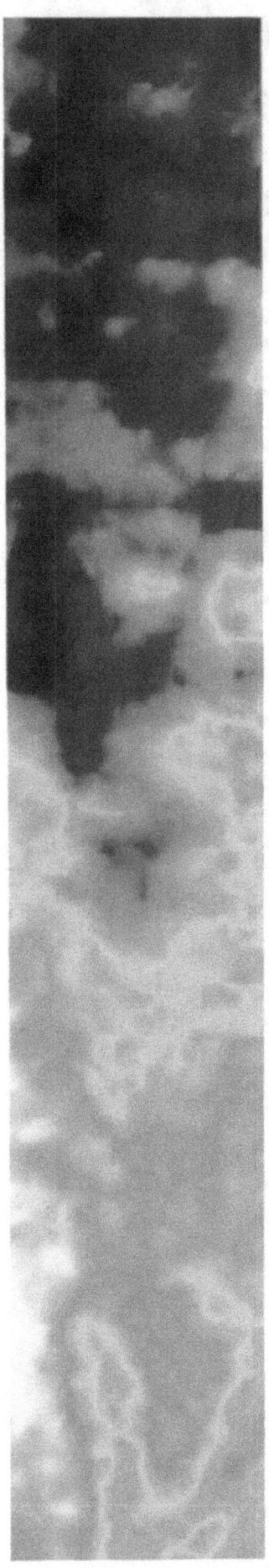
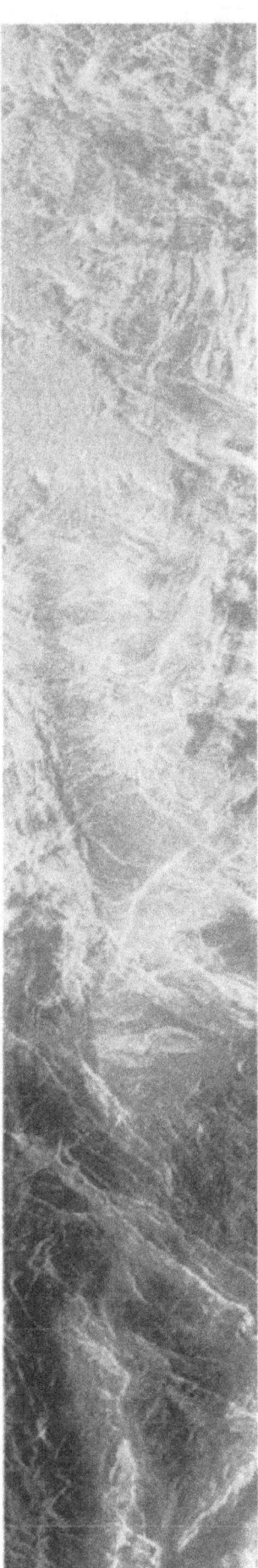

etailed view (left) of the region of highest elevations in Ovda Tessera. The image shows a return to normal radar backscatter and emissivity at the highest elevations. The change is rapid, occurring over about 200 m of altitude. The image strip at far left is a digital elevation map of the area shown in the image strip at immediate left. The elevations correspond to radii from 6055.2 km (deep blue) to 6057.1 km (red–white). The map was created from images acquired at two different viewing angles during Magellan's first and third mapping cycles. Note that the darkest area in the radar image has the highest elevation shown. 7.9

Tectonism

The surface of Earth is constantly being stretched, faulted, and folded through the motion of lithospheric plates. Venus does not show evidence of plate tectonics, but the surface of Earth's sister planet still contains a wide array of tectonic features. Their presence indicates that forces resulting from motions in Venus' fluid interior have been strong enough to cause intense folding and faulting, giving rise to mountain ranges comparable to our Himalayas and rift valleys similar to our East African Rift.

At many locations on Venus, tectonic features are associated with large centers of volcanism. Guor Linea, a major rift valley shown in perspective view, is a 45–50-km-wide, 500-km-long depression that intersects the 3-km-high volcano Gula Mons (Fig. 4.37). The rift lies on the southeastern flank of the western Eistla Regio volcanic rise, an upland formed by a large plume of hot mantle material. A second large volcano, Sif Mons, at left, rises to an elevation of 2 km. **8.1**

Forces acting to deform the surface of Venus have created numerous tectonic structures that in many places bear strong similarities to features on Earth. Over Venus' geological history, active tectonism has produced folded mountain belts, rift valleys, and an unusual, complicatedly deformed terrain, referred to as "tessera" (derived from the Greek word for "tile"), which preserves multiple episodes of both compressional as well as extensional deformation.

On Earth, the surface deformation that forms mountain ranges is driven by the relative motion of lithospheric plates. As these plates collide, stretch, or grind against each other, the resulting interactions cause intense faulting or folding. Unlike the plate tectonics of Earth, deformation on Venus is thought to be directly linked to dynamic forces within the planet's fluid mantle. Studies from gravity data suggest Venus lacks an asthenosphere — a zone of low-viscosity material that on Earth decouples the motion of the plates from flow in the mantle. On Venus, the absence of such a layer suggests that deformation at the surface can be interpreted in terms of the underlying convective motions of the planet's interior.

Tectonic deformation on Venus occurs on a variety of scales. The smallest features resolved by Magellan radar correspond to long, typically linear fractures, or faults. Within several areas, these structures are aligned in extremely uniform grid patterns. Another type of deformational feature within the plains includes low, sinuous, widely spaced ridges that strongly resemble the "wrinkle" ridges found on the Moon and Mars. The presence of the venusian ridges suggests that compression, distributed over wide regions, has caused crustal shortening.

Extensional tectonics on the terrestrial planets manifests itself in the presence of normal faults (faults with vertical offset forming steep scarps) and surface fractures. Magellan images show that this style of deformation is often concentrated in belts that lie mainly in the equatorial region and southern high latitudes of the planet. These zones are hundreds of kilometers wide and appear to form a nearly global network, often linking up with each other at large volcanic rises. Places on Venus where significant litho-spheric stretching has formed major rift valleys — depressions tens to hundreds of kilometers wide with lengths typically exceeding 1000 kilometers — are similar to sites on Earth such as the Rhine Graben, East African, and Rio Grande Rifts. Rifts on Venus are usually associated with large, domical, volcanic rises such as Beta Regio, Atla Regio, and the western part of Eistla Regio. These highlands are interpreted to be sites where large mantle plumes have caused uplift, fracturing, faulting, and volcanism.

Besides pulling it apart, lateral movement and compression have warped and buckled the crust of Venus, forming systems of valleys and ridges similar to the Appalachians and the Himalayas on Earth. The highest mountain range on Venus — Maxwell Montes in central Ishtar Terra — was formed by this process. Maxwell and the adjacent ranges of Akna and Freyja are unique to Venus' high northern latitudes in that they are found around the periphery of a large volcanic plateau (in this case, Lakshmi Planum).

Another type of landform created by the horizontal movement of the crust, found within Venus' lowland plains, consists of linear belts of closely spaced ridges. These ridge belts rise several kilometers at most above the plains and are typically hundreds of kilometers wide and thousands of kilometers long. Venus features two major concentrations of ridge belts, one at Lavinia Planitia in the high southern latitudes, and the second adjacent to Atalanta Planitia in the high northern latitudes.

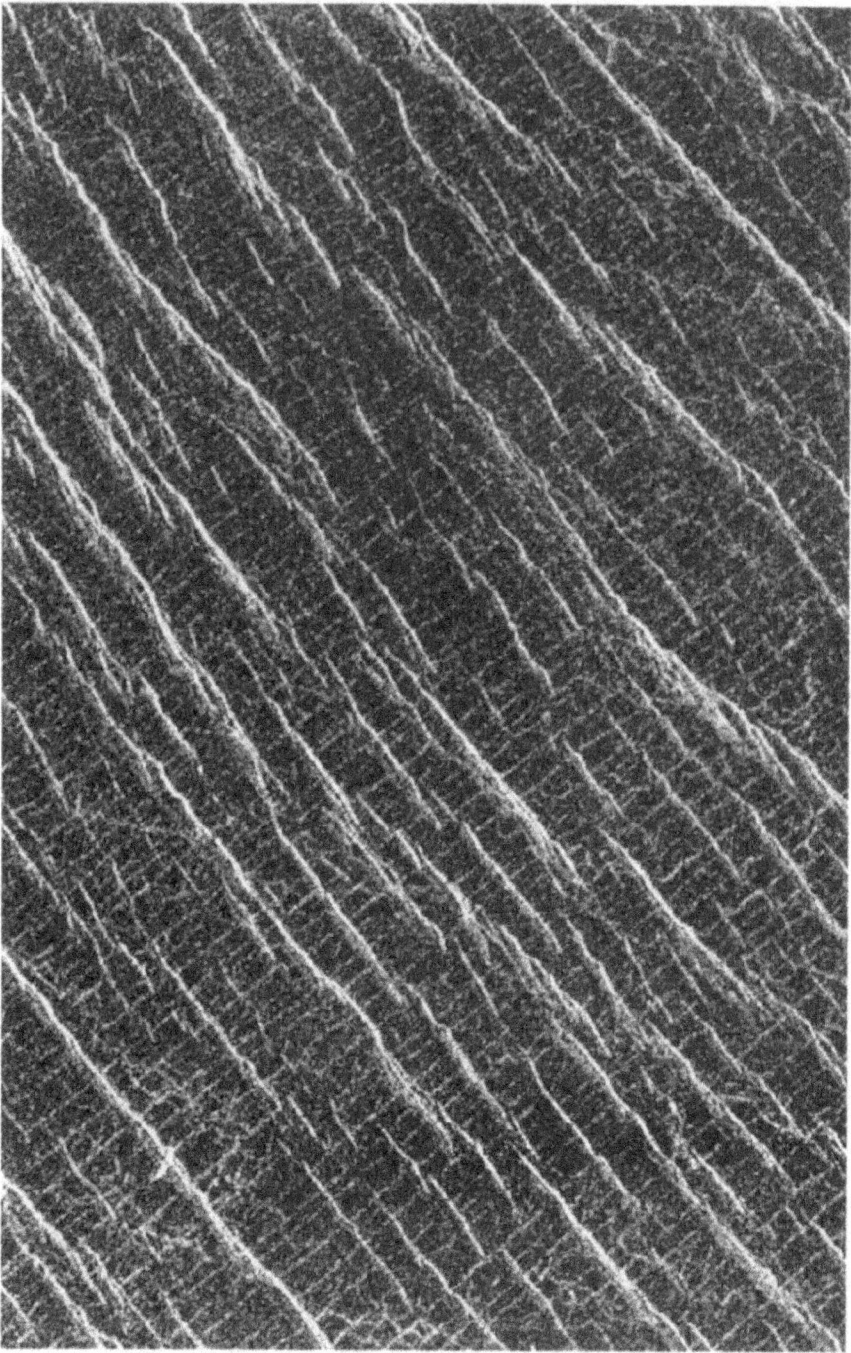

Surface deformation on Venus could be directly linked to dynamic forces in the planet's fluid mantle

This area of "gridded" plains, located in Guinevere Planitia (30.0°, 333.0°), covers an area 37 km by 80 km and is dominated by a set of northwest–southeast trending scarps. A second set of fainter lineations, with a uniform spacing of 1 km, is aligned in a northeast–southwest direction. These features are believed to have formed by stretching of the crust over wide areas. **8.2**

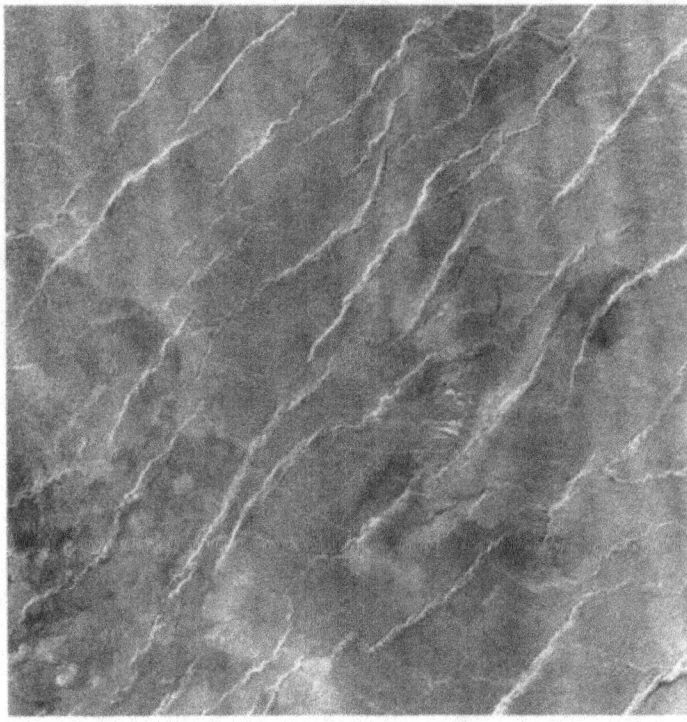

Sinuous ridges, like these in the lowlands of Helen Planitia (-34.0°, 172.0°), are found throughout the plains on Venus. They are no more than several tens of meters high and their presence suggests that modest, widely distributed, crustal compression has occurred, resulting in the formation of small folds. **8.3**

50 km

Extensional forces cause the crust of Venus to be stretched and faulted, resulting in formation of steep scarps. This group of fault scarps is located in Ganis Chasma (12.0°, 197.7°), a rift on the northern part of Atla Regio. The major scarp in the left part of the scene forms the western edge of the rift. Deformation associated with the other faults further to the east has caused a region of plains to be intensely dissected. **8.4**

12 km

The most tectonically complex terrain on Venus corresponds to high-standing areas known as complex ridged terrain, or tessera. The most conspicuous occurrences of this type of landform are at Aphrodite Terra, Alpha Regio, Tellus Regio, and the eastern part of Ishtar Terra (Fortuna). Superposition and cross-cutting relationships between different geological units indicate that these regions represent some of the oldest terrain on the planet. These highlands preserve a history marked by compression-forming ridges and troughs with later, regional-scale, extensional-deformation-producing graben cutting across the ridges. Several hypotheses to explain the formation of this terrain have been proposed. Some investigators believe that tessera may be analogous to continents on Earth. Other suggestions are that tessera may be areas where mantle downwelling has caused intense folding and faulting to produce a thick basaltic crust or sites of old mantle plumes that produced large volumes of lava on Venus' surface.

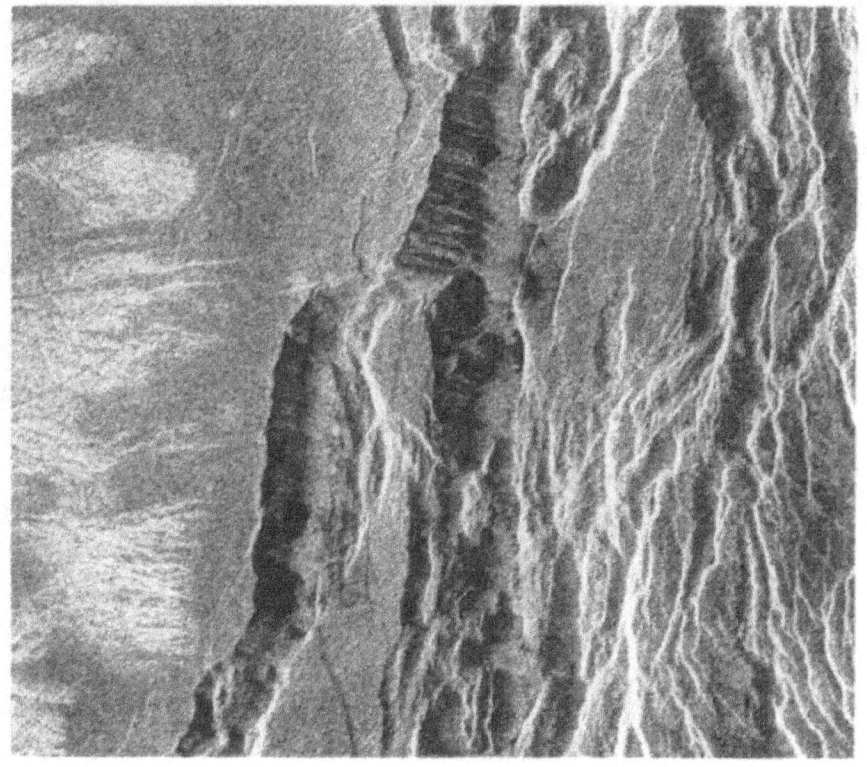

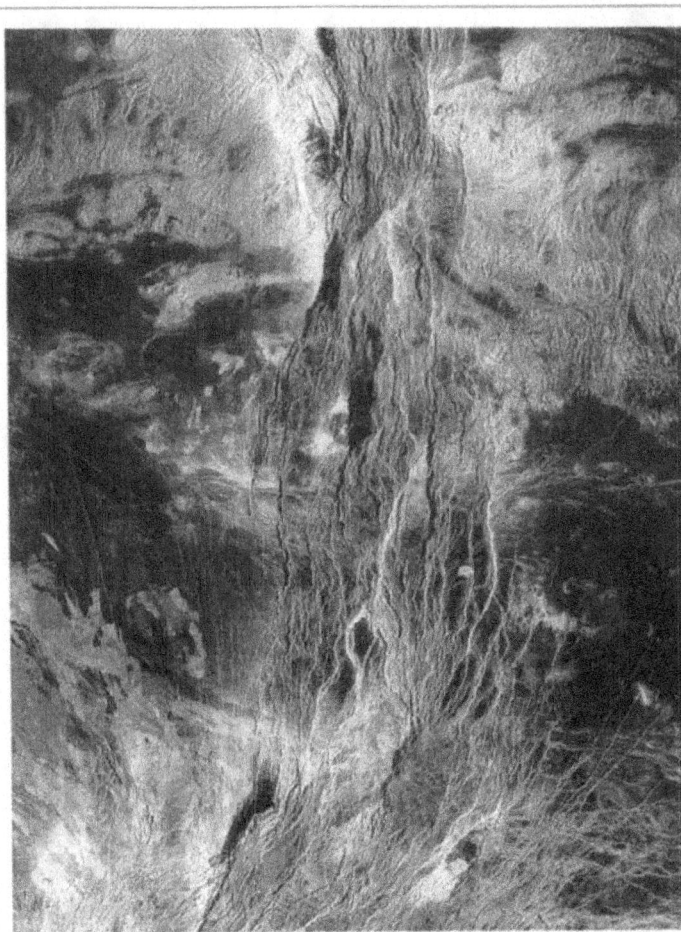

Devana Chasma, a major north–south trending rift valley, is located in the central part of Beta Regio, centered at (28.0°, 283.0°). This 1170-km-long valley is similar to areas on Earth such as the East African Rift — where the crust is being stretched and faulted. Lava flows from the volcano Theia Mons (just off the bottom left of the image) have contributed to local in-filling of the valley. In some places, more recent faulting has cut across lava flows. The 37-km-diameter impact crater Somerville, located near the center of the image (see Fig. 4.32), has been split by faulting with its eastern rim displaced by approximately 10 km. **8.5**

————— 75 km

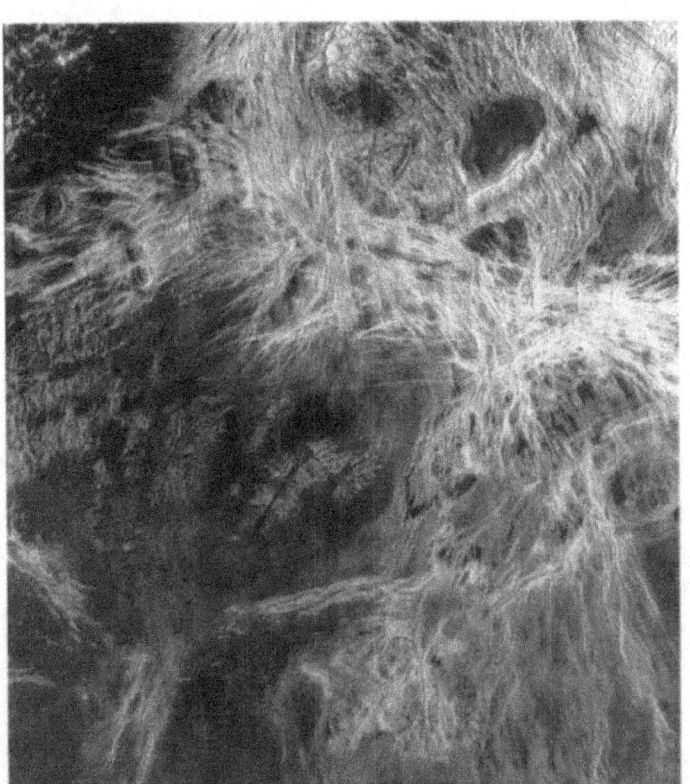

A major belt of fracturing and faulting, Ix Chel Chasma (image centered at [–12.0°, 72.5°]) strikes east–west along the southern flank of Aphrodite Terra, an elevated region of complexly deformed terrain. This zone of deformation, which is over 800 km long and 100 km wide, is part of a more globally extensive system of fracture belts. **8.6**

————— 150 km

The 5-km-high mountain range of Akna is located on the western margin of Lakshmi Planum, centered at (69.0°, 318.8°). Compressional forces, oriented in a northwest–southeast direction, have caused the crust to buckle and fold, producing this collection of northeast–southwest trending ridges and valleys. 8.7

————— 25 km

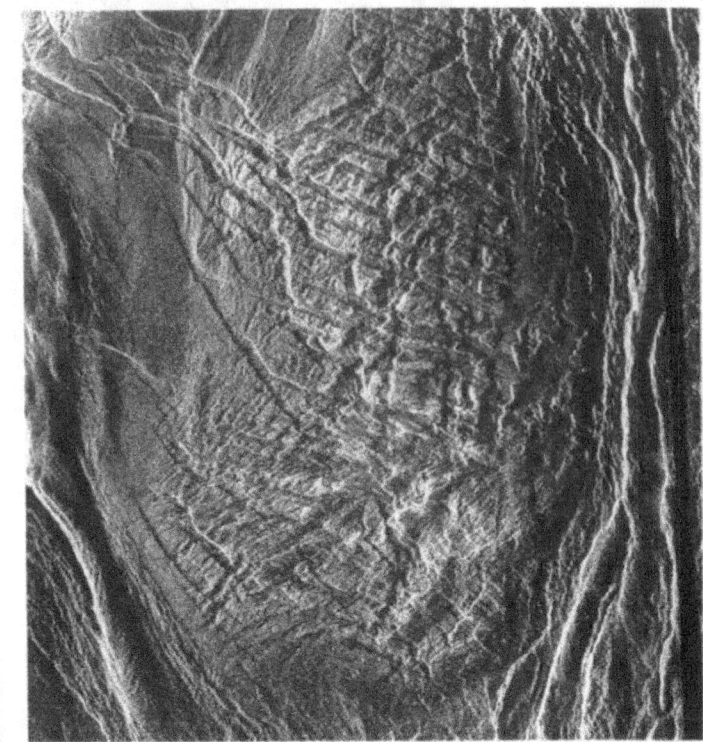

A variety of small-scale tectonic features is found within the mountain belts of Ishtar Terra. This highly fractured dome (70 km by 125 km), located on the eastern flank of Freyja Mons (72.0°, 342.0°), was formed by uplift and extension within the regionally compressional mountain belt. The fault pattern is the result of two sets of intersecting fractures, one trending roughly north to north–northeast and the other west to west–northwest. Individual scarps form graben, commonly 1–5 km wide. **8.8**

▭▭▭▭ 15 km

Maxwell Montes, centered near (65.0°, 6.0°), is the tallest mountain range on Venus, rising almost 11 km above the mean planetary radius. The western slopes of Maxwell are very steep while those to the east are more gradual. A set of north–south trending ridges and troughs was formed by lateral compression of the surface. The radar-bright appearance of this area is thought to indicate the presence of an electrically conductive mineral such as pyrite, which results in strong reflection of the incident radar signal. **8.9**

▭▭▭▭ 120 km

Pandrosos Dorsa, a ridge belt located in the lowland plains adjacent to Atalanta Planitia, was formed by general east–west oriented compression. Individual ridges within the belt are spaced 3–5 km apart with lengths that often exceed 100 km and typically form a parallel braided pattern. 8.10

60 km

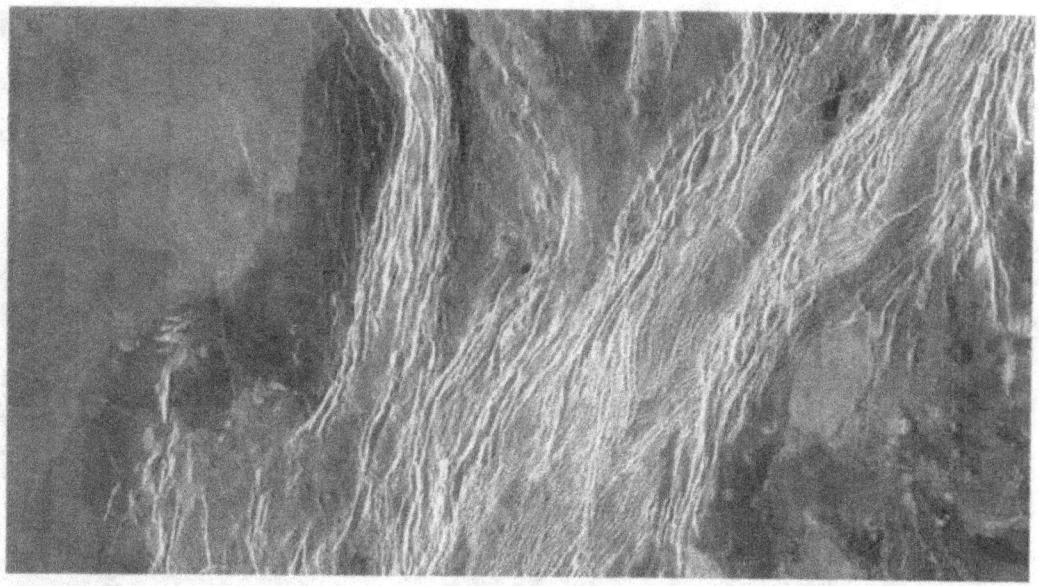

The lowlands of Lavinia Planitia contain a wide collection of features formed by compression. This scene, centered at (–45.0°, 350.0°), shows a region covering approximately 1400 km east–west and 1600 km north–south. The dominant structures are ridge and fracture belts that rise up to several kilometers above the surrounding plains and are interpreted to have formed by the compression of Venus' upper layer. Individual ridges are a few kilometers wide and tens of kilometers long. The northern part of Mylitta Fluctus, a large lava flow complex, is located at the bottom of the image. These lava deposits post-date and terminate against an elevated belt of ridges. 8.11

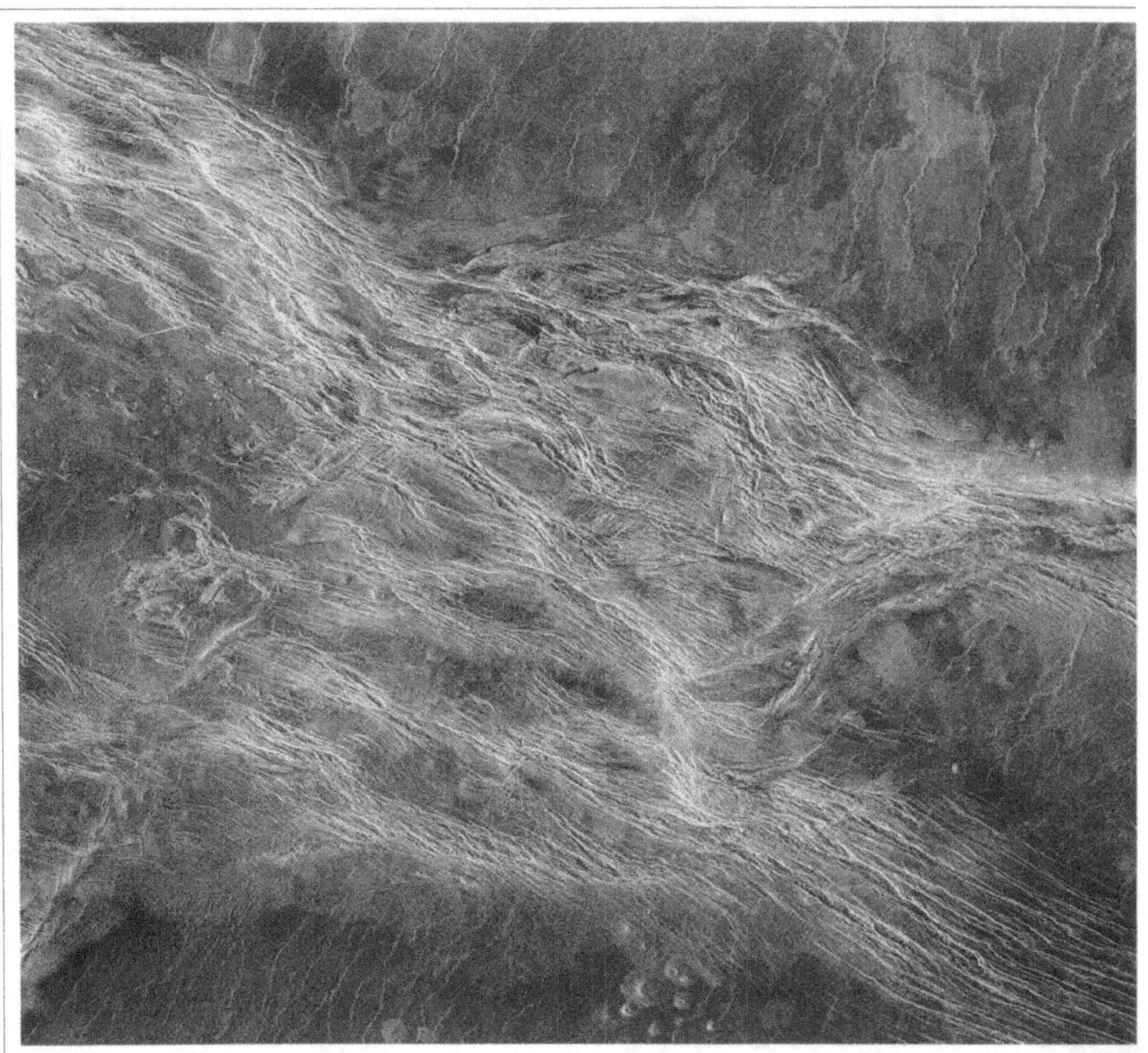

Hyppolyta Linea, a deformational belt centered at (–41.0°, 340.0°), rises 1 km above the volcanic plains of Lavinia Planitia. North–south oriented compression has caused folding and fracturing. East–west trending fractures and small graben are thought to have formed when the compressional forces caused rocks along the crest of fold structures to crack and fracture. **8.12**

━━━━━ 50 km

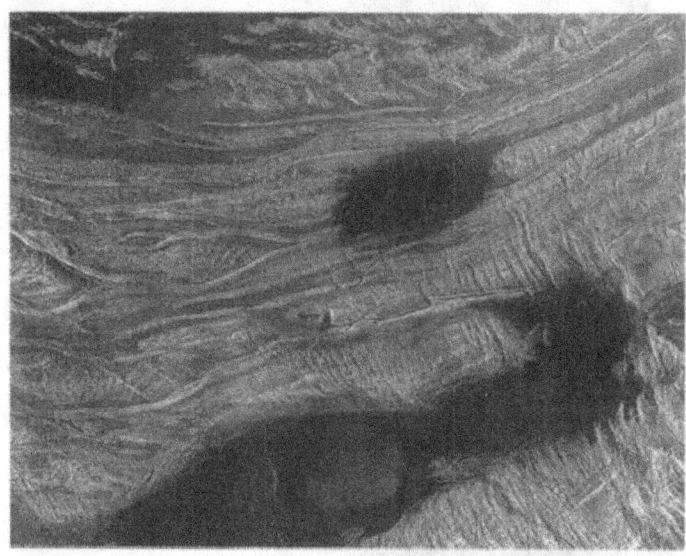

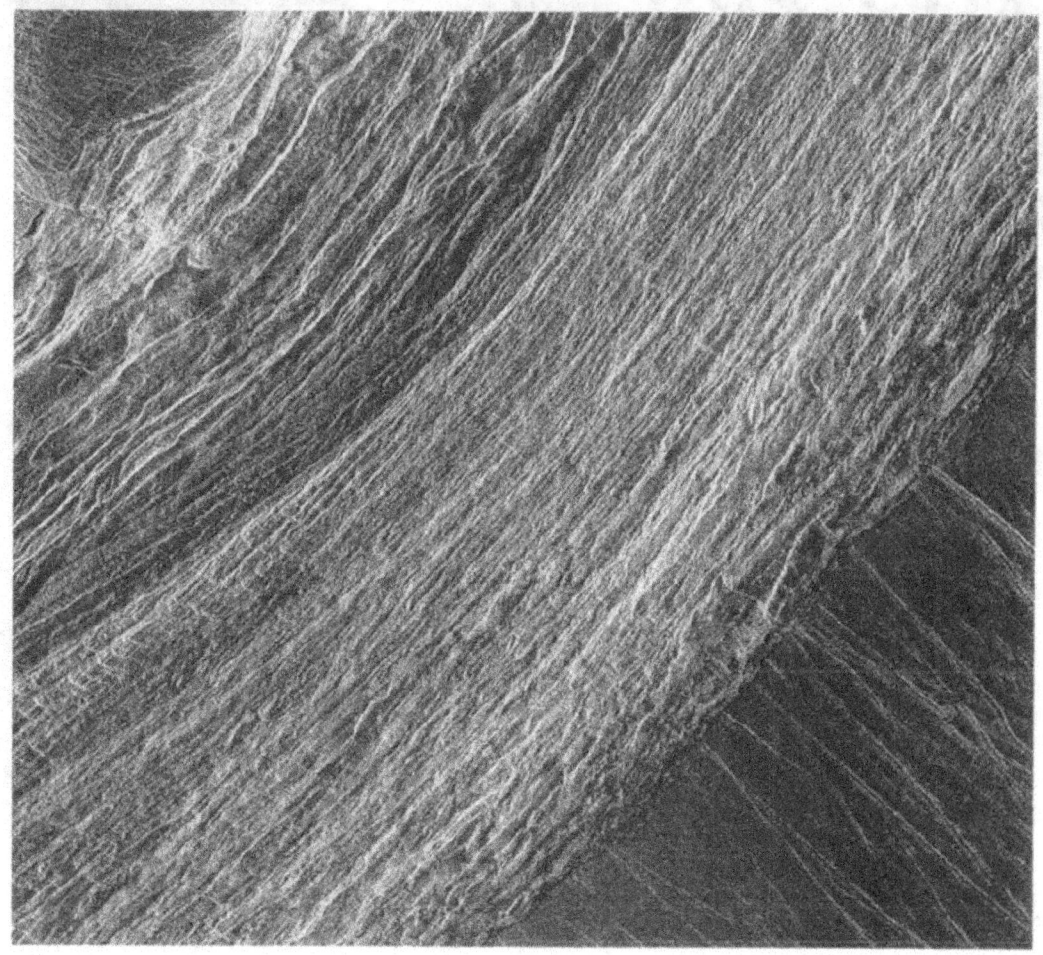

idge belts are often found along the margins of complex ridged terrain. This belt, lying along the northern edge of Ovda Regio (1.0°, 81.0°), comprises a series of east–west trending ridges 8–15 km wide and 30–60 km long. More recent north–south oriented faulting cuts across the eastern part of this area. Plains-forming volcanism has filled in low-lying areas to both north and south. **8.13**

30 km

he large corona Artemis (40.5°, 140.0°), contains structures formed by both extension and compression. The western part of this unusual arcuate feature contains steep fault scarps, suggesting that extension and normal faulting has occurred. In comparison, ridges on Artemis' eastern part appear to have formed by compression. Some investigators believe that Artemis may be a site where a process similar to subduction is occurring. **8.14**

5 kilometers

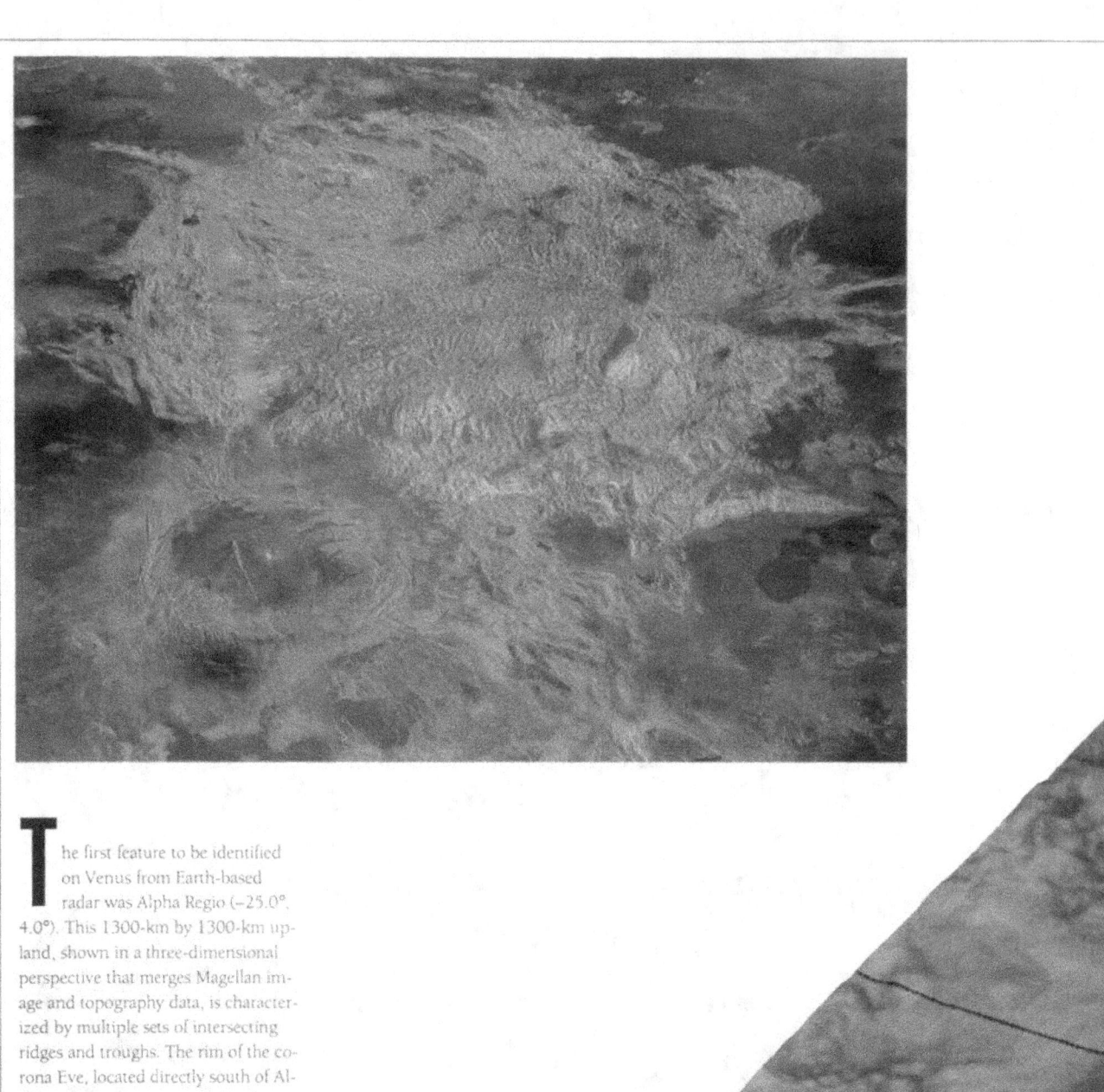

The first feature to be identified on Venus from Earth-based radar was Alpha Regio (−25.0°, 4.0°). This 1300-km by 1300-km upland, shown in a three-dimensional perspective that merges Magellan image and topography data, is characterized by multiple sets of intersecting ridges and troughs. The rim of the corona Eve, located directly south of Alpha, contains numerous faults, fractures, and graben, indicating extension has occurred. To enhance topographic detail, the vertical scale of this image has been exaggerated approximately 23 times the actual relief. **8.15**

Alpha Regio's complex topographic structure is shown in this perspective view. Color-coded elevations are overlaid on the shaded relief — the difference between the lowest and highest elevations (deep blue vs. brown) is 3 km. Elevations within Alpha Regio are quite variable, with the highest parts rising over 2 km above the surrounding plains. Many of the interior low areas are at the same elevation as the adjacent plains. Image data often show these low areas to be filled with lava deposits. 8.16

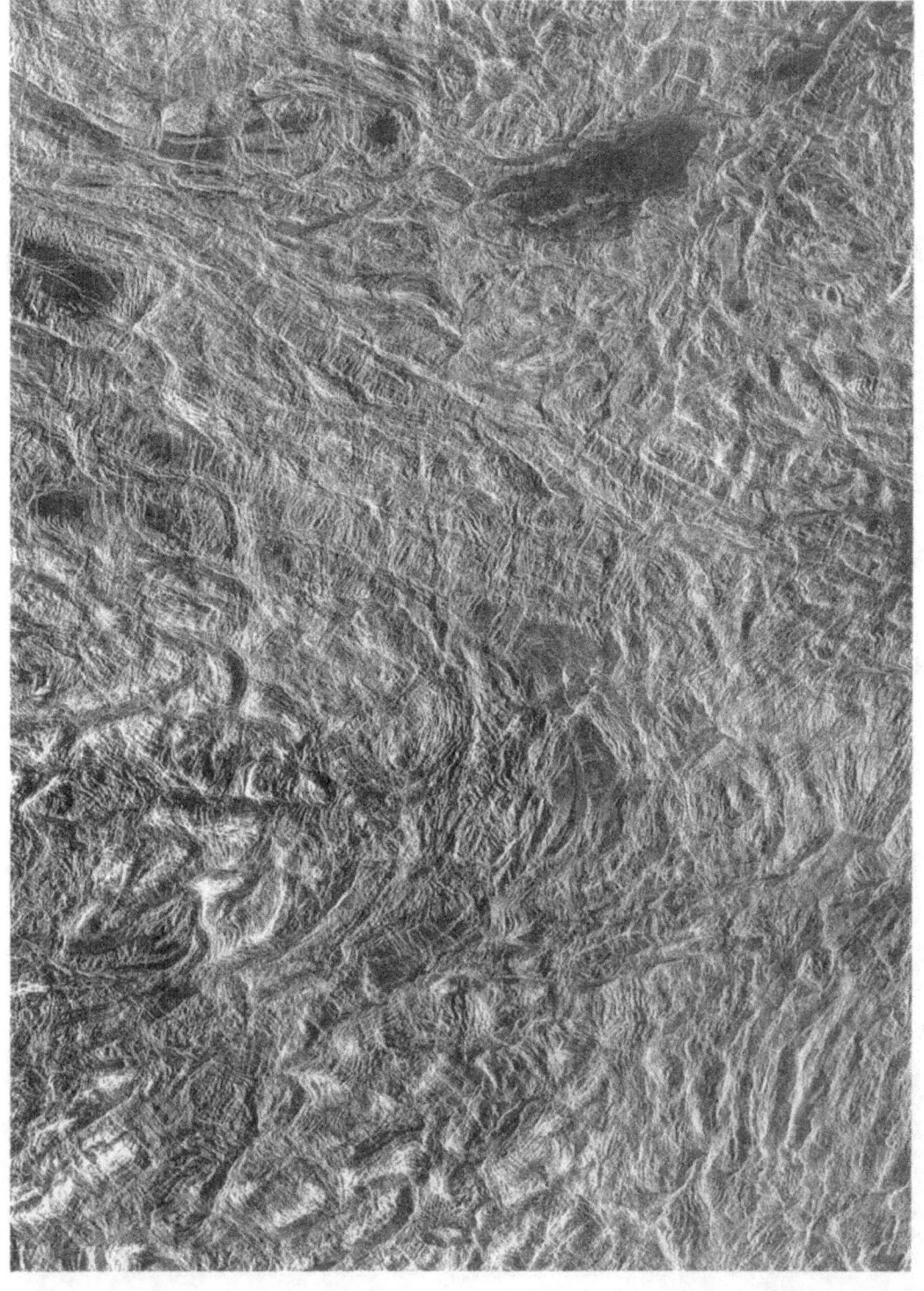

Complicated arrangement of ridges, troughs, and graben forming complex ridged terrain in the interior of Alpha Regio (–25.0°, 3.0°). In the upper left part of this scene, a set of west–northwest trending ridges and troughs, spaced 20–50 km apart, is cut by a perpendicular set of narrow graben. In the lower right of this image, a second pattern, made up of arcuate ridges and troughs with a spacing of 10–20 km, is cut by scarps, fractures, and troughs with a spacing of 1 km or less. **8.17**

50 km

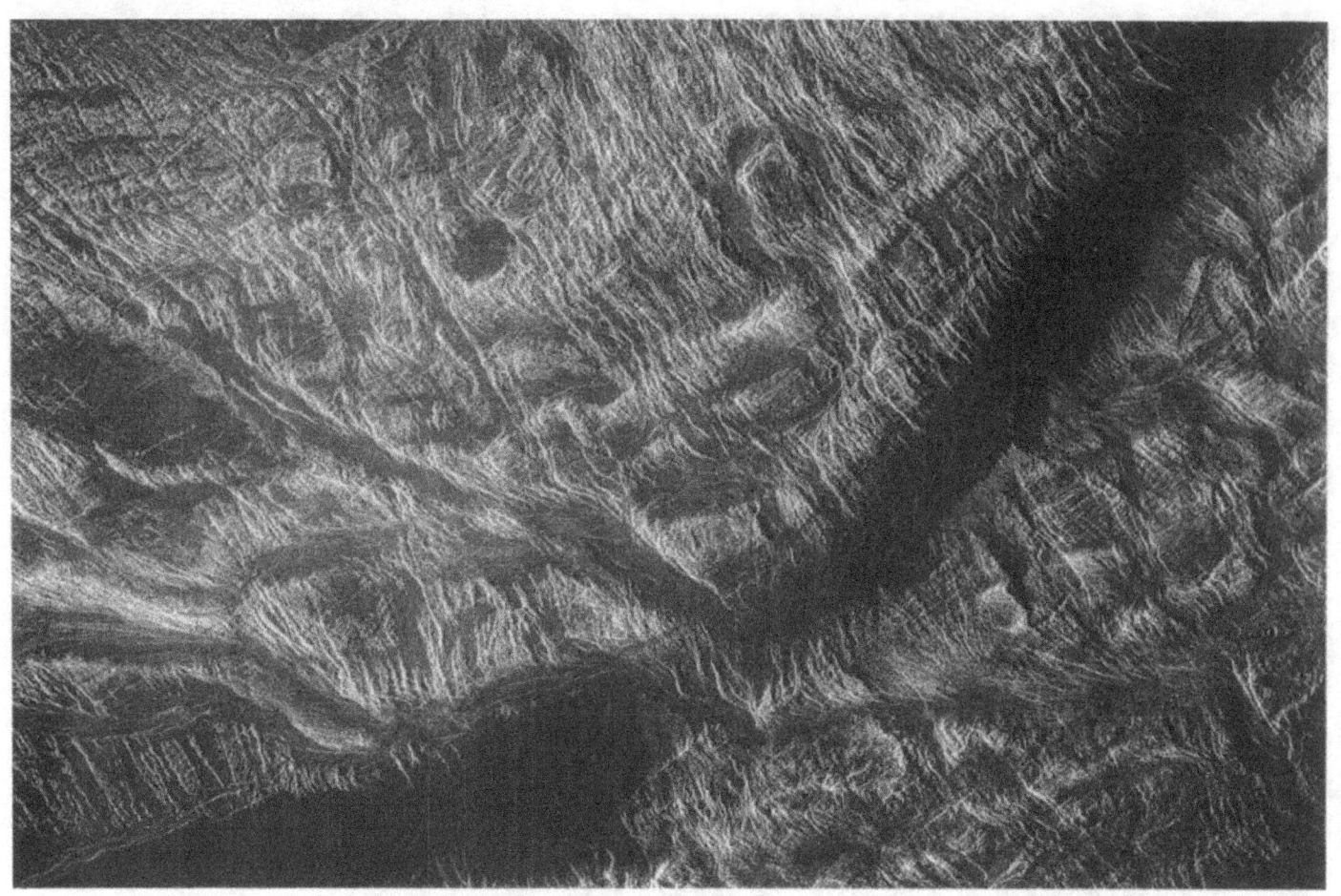

S equences of multiple episodes
of tectonic deformation resulting
in the formation of complex,
ridged terrain are shown in this part
of Ovda Regio (−1.0°, 81.0°). Within
this area, northwest–southeast oriented compression has caused folding,
resulting in the formation of a series
of northeast–southwest trending
ridges and valleys. Subsequent extension has produced northwest–southeast trending fractures and scarps.
The most recent activity corresponds
to the emplacement of dark (smooth)
lavas in low-lying areas. 8.18

———————————— 100 km

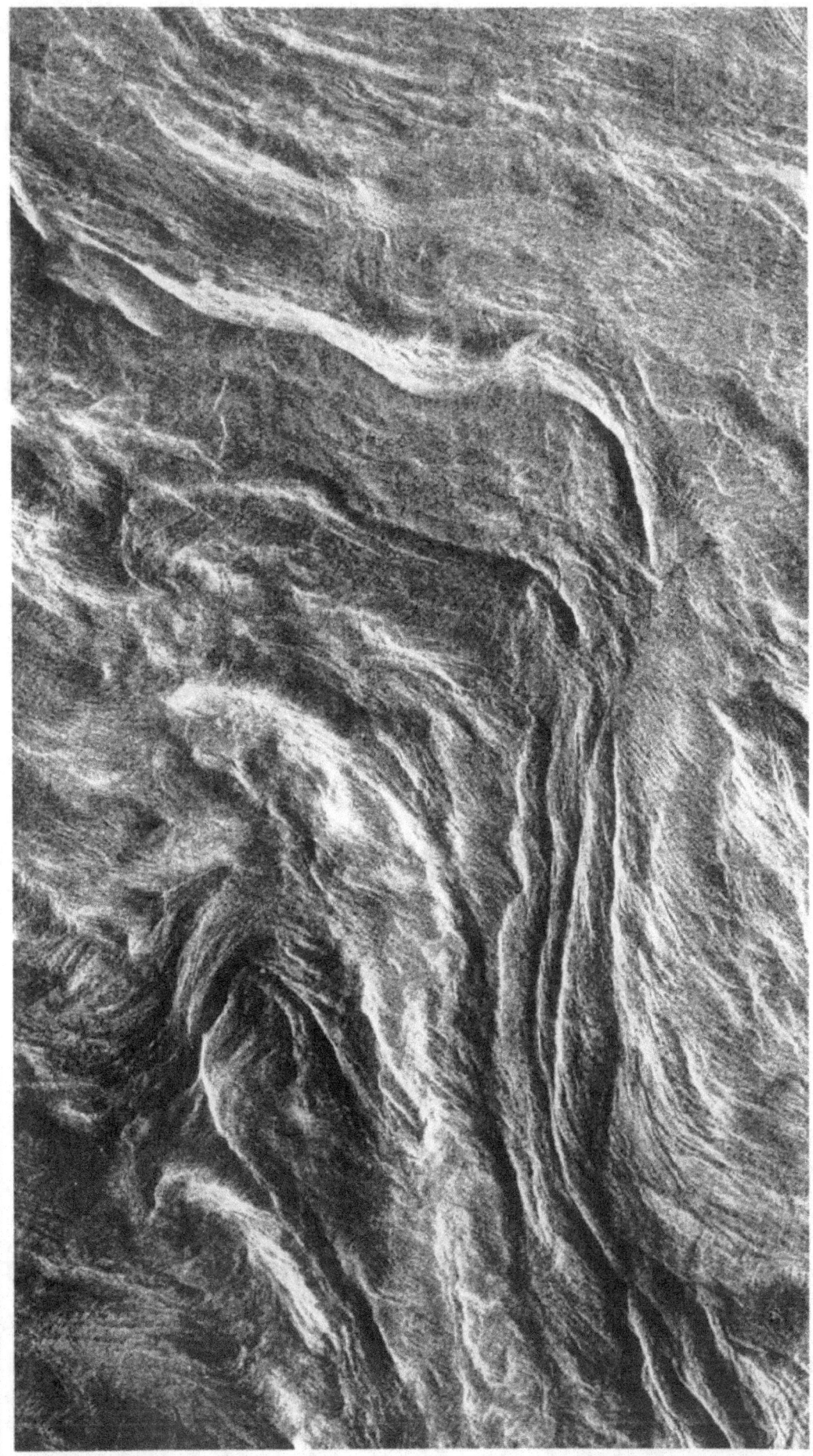

olding is a dominant process in the formation of complex, ridged terrain. This set of north–south and east–west trending ridges in the eastern part of Ovda Regio (–4.0°, 88.0°), spaced 2–5 km apart, forms a unique chevron, or V-shaped structure. Unlike other parts of Ovda, where folding is followed by extension, the absence of graben at this site is dominated by the compressional stage of tessera formation. 8.19

10 km

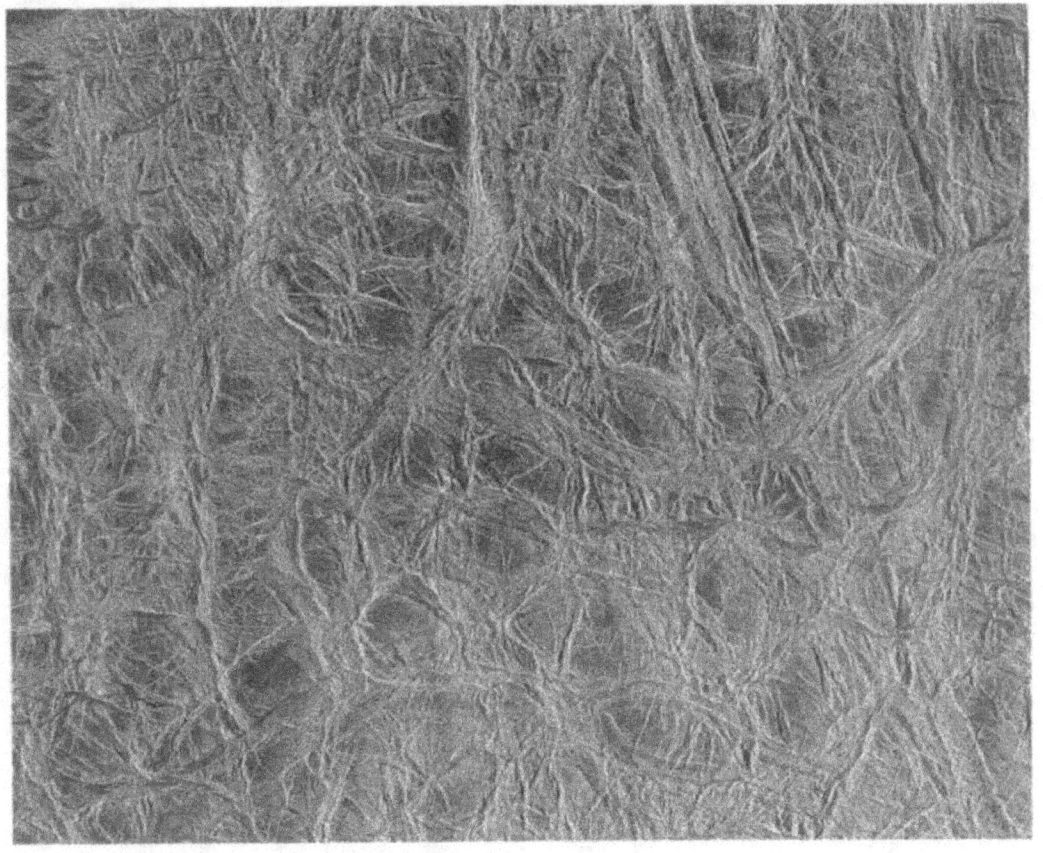

In some locations, the formation of complex, ridged terrain is dominated by extension. This portion of Phoebe Regio (−6.6°, 280.3°) contains graben that are arrayed in multiple directions and range in width from 2–15 km. **8.20**

50 km

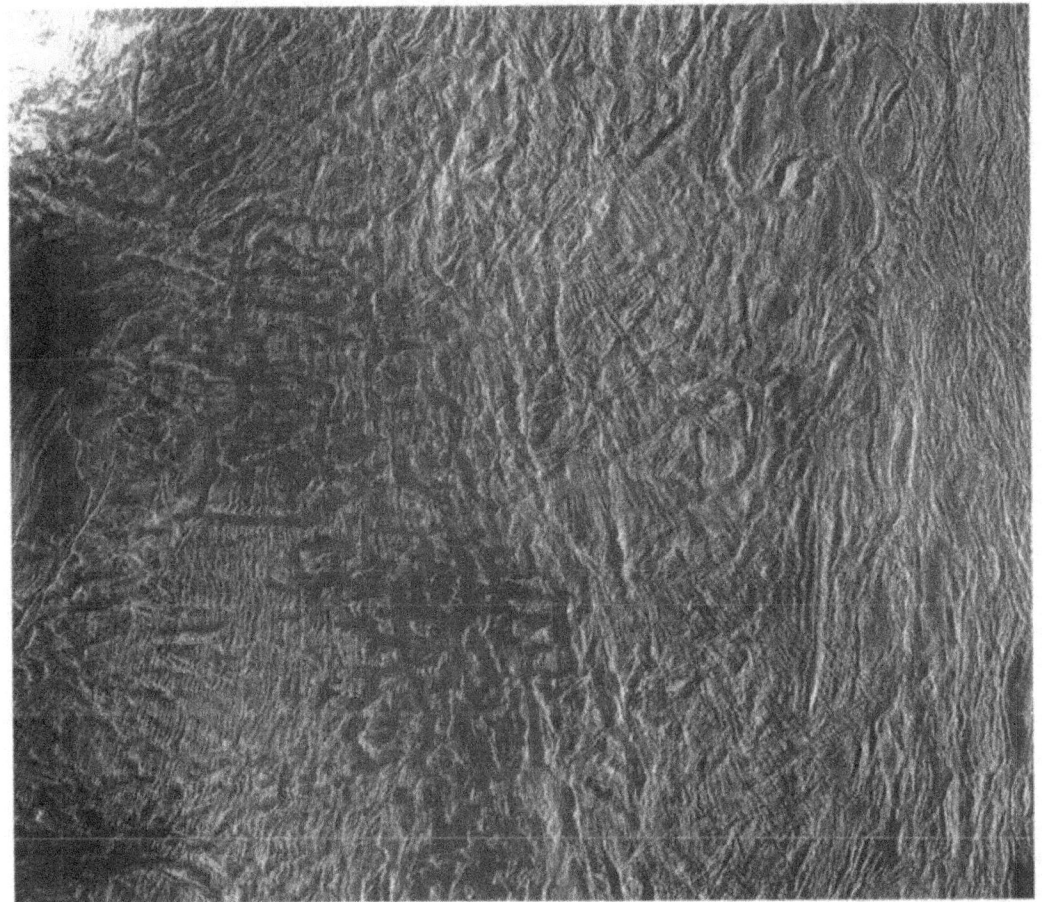

Venus' Fortuna region (60.0°, 16.0°), located to the south and east of Maxwell Montes (the radar-bright feature in the upper left of the image), contains relatively short, broad ridges, and valleys aligned in a general north–northeast direction. These structures are cut by long, north–south to north–west trending scarps and graben. In the western part of this area, more recent volcanism has filled in some of the valleys with dark, smooth, lava flows. **8.21**

100 km

Glossary

A

Advection   Transfer of heat by a lateral movement of mass.

Alluvial fan   An outspread mass of detritus deposited by a stream.

Altitude   The vertical distance of a point, not necessarily on a planet's surface, from the datum.

Anastomosing channel   A type of river channel that initially separates from its trunk stream and later rejoins the stream.

Apoapsis   In orbital motion, the point at which the orbiting body is farthest away from the primary body.

Asthenosphere   On Earth, the shell approximately 100–350 km below the surface. In this region, rock is weak and seismic waves are attenuated.

Astronomical unit   (AU or a.u.) The unit of distance in the solar system, equal to the mean distance between Earth and the Sun. The current value of the AU is 149,597,870 km.

B

Backscatter   Interaction between transmitted electromagnetic energy and its target, or the energy reflected by the target in the direction of the transmitter.

Bar   A unit of pressure, equivalent to a pressure of 0.987 atmospheres (approximately 750 mm of mercury). On Earth, atmospheric pressure of 1 bar is reached at about 100 m above mean sea level.

Basalt   Dark-colored, fine-grained rock; a solidified product of common volcanic extrusions. The silica (SiO_2) content of basalts is less than 53 percent by weight. Alkali basalts are rich in both sodium and potassium.

C

Caldera   A large, usually circular volcanic depression. Crater Lake, Oregon, formed by the eruption and subsequent collapse of Mt. Mazama, is a caldera.

Conduction   The transfer of heat by interactions on the atomic and molecular levels.

Convection   The transfer of heat by a vertical movement of mass or by a combination of vertical and lateral movements.

Core   The central portion of terrestrial planets. On Earth, the radius of the core is about 3500 km.

Crust   The uppermost layer of a terrestrial planet, composed of rock with a density lower than that of the underlying mantle. On Earth, the crustal thickness varies from about 5 km beneath oceanic basins to about 50 km beneath continental mountain chains.

D

Datum   An assumed reference level from which other levels are determined.

Dendritic drainage   A pattern of drainage channels resembling tree branches or leaf veins.

Diapir   A dome or fold created by the upward intrusion of the lighter mass into a denser, crustal rock. The most common diapiric structures on Earth have cores of salt.

Dielectric constant   Also referred to as *relative permittivity*, a quantity that characterizes the electrical properties of a substance. Since the dielectric constants of rocks are related to their composition and density, the concept of dielectric constant is used to describe the surfaces of planets that are accessible to exploration by radar.

Dike   A tabular intrusion.

E

Ejecta   Material thrown out during a volcanic eruption or crater excavation.

Elevation   The vertical distance of a point on the planet's surface, from the datum.

Emissivity   A measure of the ability of material to radiate energy produced by thermal agitation of atoms, molecules, etc.

Endogenic   Related to a geological process originating within a planet — erosion by wind or volcanism, are examples.

Eolian   Related to wind (from Aeolus, the Roman god of winds).

Exogenic   Related to a geological process originating outside a planet — meteoroids, for instance.

F

Fault   A fracture in rock, along which observable displacement has occurred.

Ferroelectrics   Substances that exhibit electrical properties, such as dielectric hysteresis, that are analogs to certain properties of ferromagnetic substances.

Fold   A bend or flexure in the rock strata.

G

Graben   A depressed crustal block, bounded by faults on its long sides.

Gravity anomaly   The difference between the observed and calculated values of acceleration of gravity at a given location. The free-air anomaly ignores the effects of topography and isostatic compensation

H

Hadley cell   Global atmospheric circulation pattern in which the upward motion of air over the tropical latitudes is compensated by the downward motion at higher latitudes. The resulting trade winds tend to blow toward the equator.

Hydrazine   A high-energy fuel (N_2H_4), used as a rocket propellant

I

Impact crater   A terrain depression, usually circular in outline, created by a collision with a body in interplanetary orbit. In size, impact craters can range from microscopic pits, found in lunar rocks, to major topographic cavities such as the Hellas basin on Mars or Mare Imbrium on the Moon.

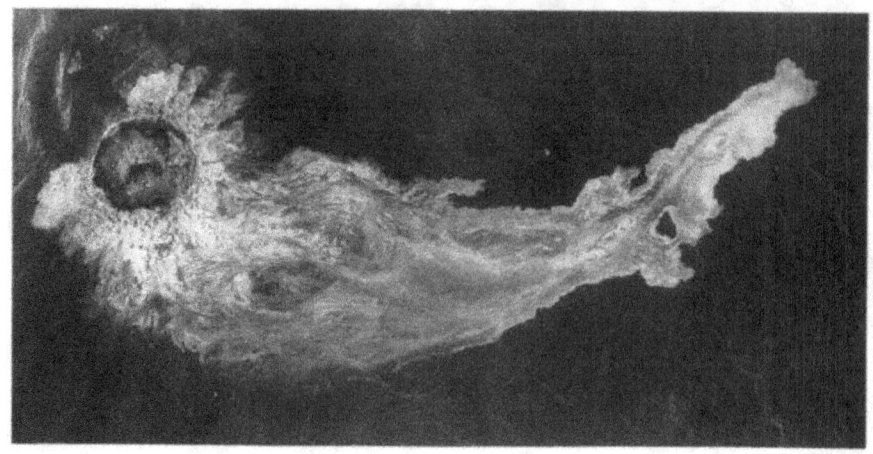

(4.27)

Incidence angle   The angle between the radar beam and the normal to the local horizontal. Radars designed for geological imaging usually operate at incidence angles from 20° to 50°.

Inferior conjunction   The position of either Venus or Mercury on the line between Earth and the Sun.

Inferior planets   Venus and Mercury, also referred to as the *inner planets*.

Intrusion   A body of igneous (crystallized from a magma) rock, emplaced within the pre-existing rock.

Isostasy   Condition of equilibrium in Earth's crust.

K

Kelvin   Unit for measuring temperature. The kelvin scale starts at *absolute zero* (−273.16° Celsius), with each kelvin corresponding to an interval of one degree Celsius.

L

Lineament   A linear topographic feature of regional extent, believed to be an expression of the underlying crustal structure. Examples include fault lines, valleys controlled by faulting, the straight fronts of mountain ranges, linearly aligned hills or volcanoes, etc.

Lithosphere   The brittle, uppermost layer of Earth's crust, on the average about 100 km in thickness.

M

Mantle   In terrestrial planets, the spherical shell between the core and the crust. On Earth, the mantle extends to a depth of about 2900 km.

Mantle convection   Mass movement hypothesized to be taking place in the Earth's mantle and perhaps also in the mantles of other terrestrial planets. The movement is thought to be primarily driven by radiogenically released heat

Mantle plume Melted, buoyant rock believed to be rising through the mantle.

Mass wasting General term for transport of rock debris, down slope, under the influence of gravity.

Meridian The great circle on the celestial sphere joining the celestial poles through the zenith.

Multiring basin A large impact crater, surrounded by a progression of impact-generated concentric ridges and troughs. An example is the Mare Orientale basin on the Moon.

N

Nadir The point directly below the observer.

Normal fault A fault in which the hanging wall has moved or appears to have moved downward.

P

Patera Usually a volcanic caldera with particular morphological characteristics. As a distinct landform type, paterae were first identified on Mars.

Periapsis In orbital motion, the point at which the orbiting body is closest to the primary body.

Perovskite The mineral $CaTiO_3$.

Plate A rigid section of the lithosphere that floats on the underlying asthenosphere.

Primary crater A crater excavated in the location of contact between the projectile and the target. Secondary craters are excavated by the debris ejected in the course of excavation of the primary crater.

Pyrite A common sulphide mineral (FeS_2), also known as "fool's gold" for its metallic luster.

Pyrochlore An ore for niobium, tantalum, and the elements in the rare earth group.

R

Reaction wheels A system of electrically driven fly-wheels, with axes usually along mutually orthogonal directions, that are used for controlling spacecraft attitude and pointing.

Reflectivity A measure of the ability of material to reflect electromagnetic energy. Reflectivity and emissivity are complementary quantities.

Rift A fracture in rock that resulted from one block slipping past another, usually in parallel with the regional structural trends.

Rift valley A trough that has developed along a rift; often synonymous with graben. A major example of a rift valley on Earth is the East African rift system.

Root-mean-square value (RMS or rms value). The effective value.

S

Scarp A steep hill-slope produced by erosion or faulting.

Solid rocket motor A rocket engine consisting of two major components: the combustion chamber containing the propellant, and the nozzle. The propellant comprises a solid mixture of fuel and oxidizer.

Stratigraphy The study of rock strata and their interpretation in the context of geological history. Also, the arrangement of rock strata.

Subduction A plate-tectonics process wherein one lithospheric plate descends beneath another plate. On Earth, oceanic trenches along the continental margins are associated with subduction.

Synthetic aperture radar (SAR) A technique in which a radar image with resolution corresponding to that obtained by a large, stationary antenna is synthesized using the signals transmitted and received by a small antenna mounted on a moving platform.

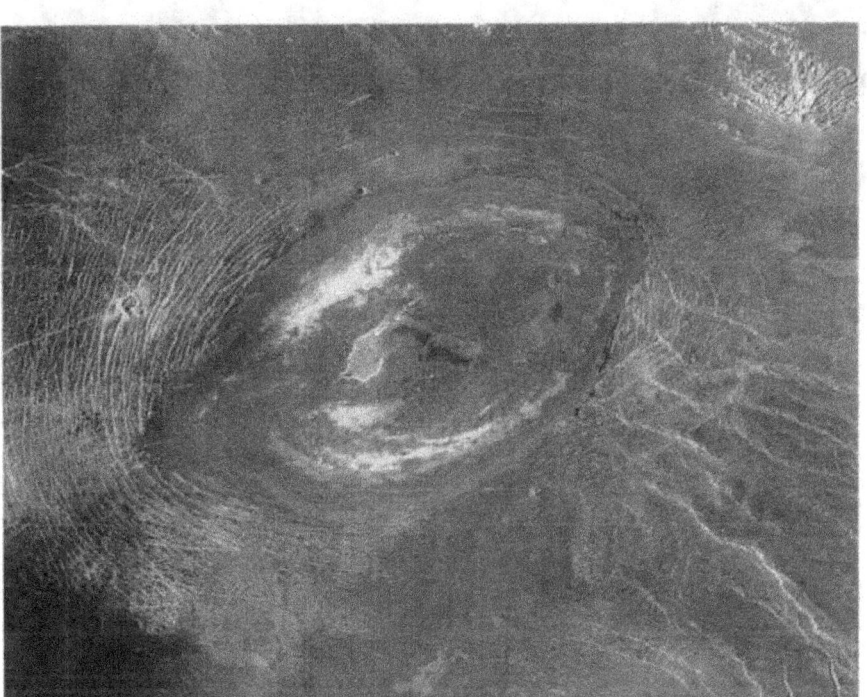

(5.8)

T

Tectonic   Related to the rock structure which resulted from deformation of the planet's outermost layers. The term does not necessarily specify the origin of deformation.

Tectonics   A branch of geology dealing with the major structural features of a planet's outermost layers. In the theory of plate tectonics, Earth's lithosphere is divided into a number of plates that move in relation to each other.

Terminator   The boundary between the illuminated and dark sides of a planet.

Terrestrial planets   Mercury, Venus, Earth, and Mars — the rocky planets of the inner solar system. The Moon is often included among the terrestrial planets. The term *inner planets* is also used.

Transfer orbit   In interplanetary travel, an elliptical trajectory tangent to the circumsolar orbits of both the departure planet (Earth) and the target planet (Venus, for instance).

Transit   The passage of either Venus or Mercury across the disk of the Sun. Also, the passage of a celestial object across the observer's meridian.

V

Volcanic shield   A broad volcanic cone, usually of a large areal extent. The volcanoes Kilauea and Mauna Loa on the island of Hawaii are examples of volcanic shields.

W

Wrinkle ridge   A topographic form first observed on the Moon, consisting of elongated segments of positive relief. The lunar wrinkle ridges may be tens of kilometers long and several tens of meters high.

X

X-ray fluoroscence   A method of chemical analysis used to identify the presence of heavy elements.

Y

Yardang   A wind-sculpted topographic form.

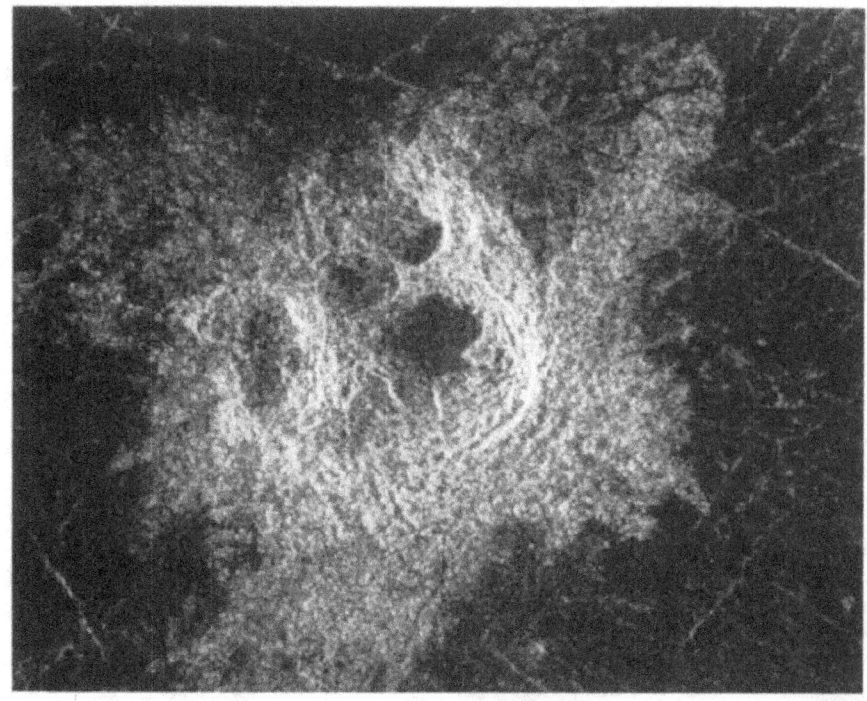

(4.3)

Nomenclature of Venusian Features

T his appendix lists the names of all the venusian surface features mentioned in *The Face of Venus* and provides a brief explanation of their origin. Naming conventions in astronomy and planetary science fall under the jurisdiction of the International Astronomical Union (IAU). In keeping with traditional association of the planet Venus with feminine symbols, the IAU's Working Group on Planetary System Nomenclature and its Task Group on Venus Nomenclature decided to assign female names to the planet's topographic features. Large craters and volcanic paterae were named after distinguished historical figures. Female first names were used to identify the smaller craters. Features of non-impact origin were assigned names of female mythological characters.

The names Alpha, Beta, and Maxwell, although not female, were incorporated in the venusian nomenclature. All three designations refer to the bright features discovered in the first Earth-based radar images of Venus (Figures 1.6, 1.7, and 1.9). The designation Maxwell honors the physicist whose contributions to the development of the theory of electromagnetic field made radio and radar possible. For more details on the nomenclature of venusian surface features, see, e.g., Fimmel et al., (1983), Burba (1990), or Greeley and Batson (1990).

A

Addams   Jane Addams (1860–1935), American social reformer

Adivar   Halide Adivar (1883–1964), Turkish educator

Aglaonice   Greek astronomer

Aino   In Finnish mythology, a water spirit

Akna   In Yucatán mythology, goddess of birth

Alpha   First letter of the Greek alphabet

Annia Faustina   Wife of Roman emperor Marcus Aurelius

Aphrodite   In Greek mythology, goddess of love

Ariadne   Greek female first name

Artemis   In Greek mythology, goddess of the Moon and hunt

Asteria   In Greek mythology, a Titaness

Astrid   Scandinavian female first name

Atalanta   In Greek mythology, huntress associated with golden apples

Atla   In Norse mythology, a giantess and the mother of Heimdall

Atropos   In Greek mythology, one of the three Fates

Aurelia   Mother of Roman emperor Julius Caesar

B

Barton   Clara Barton (1821–1912), founder of the American Red Cross

Bereghinya   In Russian mythology, a water spirit

Beta   Second letter of the Greek alphabet

Buck   Pearl S. Buck (1892–1973), American writer and Nobel laureate

C

Carson   Rachel Carson (1907–1964), American biologist

Cleopatra   (69–30 B.C.), Queen of Egypt

Clotho   In Greek mythology, one of the three Fates

Colette   Claudine Colette (1873–1954), French writer

Cunitz   Maria Cunitz (1610–1664), Polish astronomer and mathematician

D

Dali   In Georgian mythology, a goddess

Danilova   Maria Danilova (b. 1793), Russian dancer

Danu   In Celtic mythology, a goddess

Devana   In Czech mythology, goddess of the hunt

Diana   In Roman mythology, goddess of the Moon

Dickinson   Emily Dickinson (1830–1886), American poet

E

Eistla   In Norse mythology, a giantess

F

Fortuna   In Roman mythology, goddess of chance

Freyja   In Norse mythology, a goddess and mother of Odin

G

Ganiki   In Orochian (Siberian) mythology, a mermaid

Ganis   In Lapp mythology, a forest maiden

Golubkina   Anna Golubkina (1864–1927), Russian sculptor

Guinevere   In English mythology, wife of King Arthur

Gula   In Babylonian mythology, an Earth mother

Guor   In Norse mythology, a Valkyrie

H

Hathor   In Egyptian mythology, goddess of the sky

Helen   Greek female first name

Heng-O   In Chinese mythology, goddess of the Moon

Hestia   In Greek mythology, goddess of the hearth

Hippolyta   In Greek mythology, Queen of Amazons

I

Imdr   In Norse mythology, a giantess

Innini   In Babylonian mythology, an Earth mother

Isabella   (1451–1504), Queen of Spain

Ishtar   In Babylonian mythology, goddess of love

Ix Chel   In Aztec mythology, wife of the Sun god

J

Jeanne   French female first name

Juno   In Roman mythology, sister of Jupiter

K

Kawelu   In Hawaiian mythology, a heroine

Klenova   Maria Klenova (1898–1976), Soviet marine biologist

L

Lada   In Slavic mythology, goddess of love

Lakshmi   In Hindu mythology, goddess of prosperity

Lavinia   In Roman mythology, wife of Aeneas

Lilian   female first name

Lydia   female first name

M

Maat   In Egyptian mythology, goddess of truth and justice

Markham   Beryl Markham (1902–1986), English aviatrix

Maxwell   James Clerk Maxwell (1831–1879), Scottish physicist

Mead   Margaret Mead (1901–1978), American anthropologist

Mylitta   In Semitic mythology, a mother goddess

N

Navka   In Arabic mythology, a mother goddess

Neyterkob   In Masai mythology, goddess of Earth and fertility

Niobe   In Greek mythology, the first mortal with whom Zeus mated

Nokomis In Algonquin mythology, an Earth mother

O

Onatah In Iroquois mythology, a corn spirit

Ovda In Volga River–region mythology, a mistress of supernatural power

P

Pandrosos In Greek mythology, the first woman to spin a yarn

Phoebe In Greek mythology, a Titaness

Q

Quilla In Incan mythology, a Moon goddess

R

Rhea In Greek mythology, a Titaness

Riley Margaretta Riley (1804–1899), English botanist

(4.13)

S

Sabira Tatar female first name

Sacajawea (1788–1812? or 1884?), Shoshone woman who accompanied the Lewis and Clark expedition to the Pacific Northwest

Sachs Nelly Sachs (1891–1950), Swedish playwright

Sapas In Phoenician mythology, a goddess

Saskia Wife of Rembrandt

Sedna In Eskimo mythology, goddess of the sea

Sedna In Eskimo mythology, goddess of the sea

Sekmet In Egyptian mythology, goddess of war

Sif In Norse mythology, a goddess and wife of Thor

Somerville Edith Somerville (1862–1915), Irish writer

Stein Gertrude Stein (1874–1946), American writer

Stephania female first name

Stuart Mary Stuart (1542–1587), Queen of Scotland

T

Tellus In Roman mythology, goddess of Earth

Tethus In Greek mythology, a Titaness

Theia In Greek mythology, a Titaness

Themis In Greek mythology, a Titaness

U

Ulfrun In Norse mythology, a giantess

Ushas In Hindu mythology, goddess of dawn

V

Vinmara In New Hebridean mythology, a swan maiden

Vellamo In Karelo-Finnish mythology, a mermaid

Z

Zamudio Adela Zamudio (1854–1928), Bolivian poet

Identification of Magellan Images

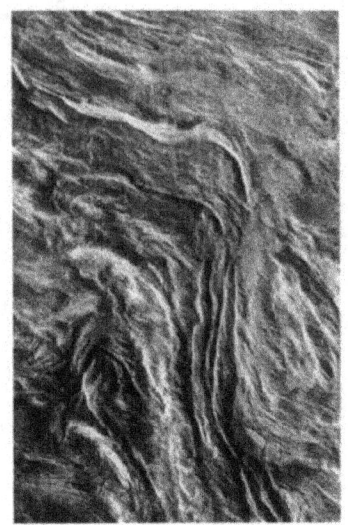

(8.19)

The images reproduced in the *The Face of Venus* constitute but a small subset of the volume of data produced during the Magellan mission. Handling, processing, and archiving these data represented a major endeavor for Magellan project and science investigators — and for various NASA elements, including the Jet Propulsion Laboratory, the Deep Space Network, and the Planetary Data System. In addition to the list of images given in this appendix, the following paragraph provides some rudimentary information on Magellan data products. For detailed descriptions of these products, please see Saunders, et al (1990); Ford, et al (1993); and Yewell (1990).

Radar-imaging data acquired during each orbit were processed into a sinusoidal equal-area format to yield image strips called Full-resolution Basic Image Data Records (F-BIDRs). Each F-BIDR strip is approximately 350 picture elements (pixels) wide and well over 200,000 pixels long. For Venus' polar areas, separate Full-resolution Polar Image Data Records (F-PIDRs) were compiled. To make the imaging data more accessible — since the full-resolution strips in raw form are too unwieldy for scientific analysis — segments of contiguous strips were processed to form mosaics of parcels of the venusian surface bounded in longitude and latitude (lon × lat). These mosaics are called Mosaicked Image Data Records (MIDRs). In coverage and resolution, MIDRs range from Full-resolution Mosaicked Image Data Records (F-MIDRs, 5° × 5°); through once-compressed mosaics (C1-MIDRs, 15° × 15°); through twice-compressed mosaics (C2-MIDRs, 45° × 45°); to thrice-compressed mosaics (C3-MIDRs, 120° × 80°). In each compression step, the linear resolution (size of pixels) is degraded by a factor of three. In the symbols xxNxxx or xxSxxx, adopted for the purpose of identifying the MIDR mosaics, the first two digits give the latitude (north or south), and the last three digits give the longitude of the center point. The MIDR mosaics form the standard Magellan data products, which are available on compact discs with read-only memory (CD-ROM). Images identified in this listing with a "MRPS" number are referenced by a JPL negative number. Images identified with a "P" number have been released by the JPL Public Information Office.

CHAPTER 1

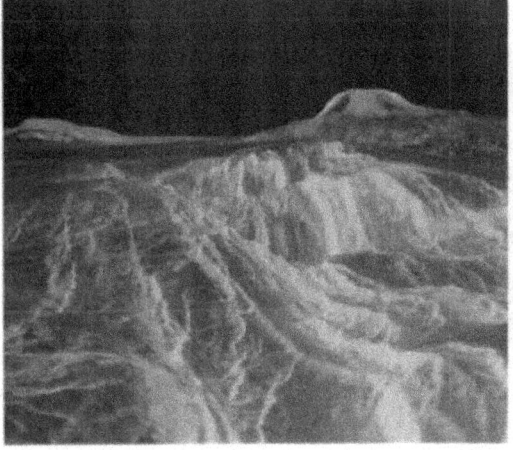

(8.1)

CHAPTER 5

CHAPTER 6

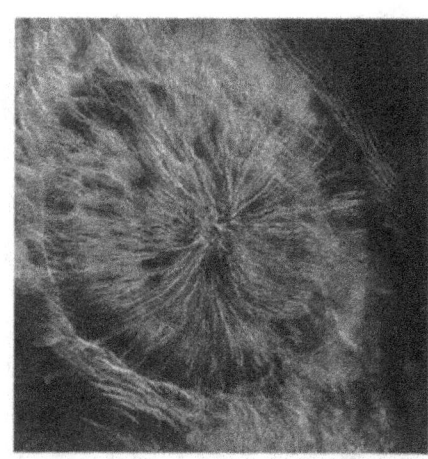

(5.23)

(8.14)

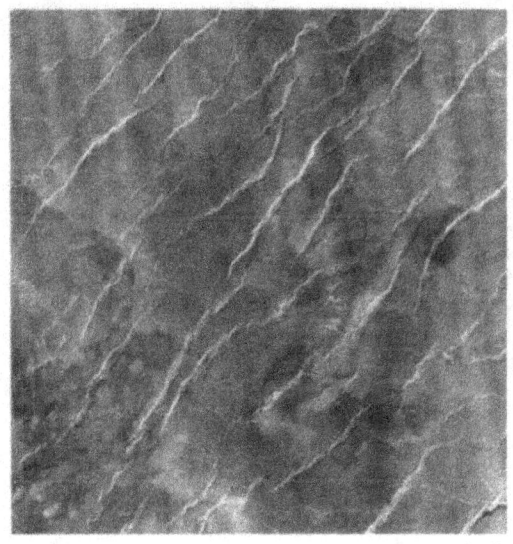

(8.3)

Availability of Magellan Data Products

F or additional information about the Magellan mission, for access to Magellan imagery or data or for detailed catalog information, refer to the following sources (from Okerson [1992]; see also Ford, et al [1993]).

NATIONAL SPACE SCIENCE DATA CENTER

The National Space Science Data Center (NSSDC) at NASA's Goddard Space Flight Center is the principal archive and distribution center for all NASA missions. The center contains all the standard mosaic image products released by the Magellan project in both photographic and digital form on compact-disc read-only memory (CD-ROM) format. The center also makes available press-release images, videotapes, software for displaying the CD-ROM digital images, planning maps of Venus, a fact sheet and other documentation.

NSSDC's principal charter is to support data distribution to researchers. Requests are filled (without charge) from NASA centers, federal, state and local governments and NASA-funded researchers. Requests from others are filled for a nominal charge, but the center may waive the fee — in the case of educational requests for a limited quantity of material.

NSSDC has the equipment necessary to supply special requests such as large photographic enlargements. The center has a limited staff that is able to assist with questions about data products or the identification of products showing specific features. The center cannot, however, support requests for extensive assistance from researchers, the public or teachers.

A general catalog showing the types of materials available from NSSDC is available for online computer access. Requests for materials can be made through the online catalog, by electronic mail (E-mail) or by telephone. Contact the center as follows:

National Space Science Data Center
Goddard Space Flight Center
Greenbelt, Maryland 20771
Tel: (301) 286-6695
Fax: (301) 286-4952

Online Catalog
NSSDC::
or
nssdc.gsfc.nasa.gov
(128.183.36.25)
Username: nssdc

E-mail
NSSDC::REQUEST
or
request@nssdc.gsfc.nasa.gov

PLANETARY DATA SYSTEM

Researchers funded by NASA can obtain Magellan materials through the Planetary Data System (PDS). The system consists of a central, online catalog at JPL — and a number of "nodes" located at research facilities with particular expertise in specific planetary research areas. The PDS Geosciences Node at Washington University is particularly responsible for cataloging and supporting Magellan data.

The Geosciences Node provides a direct, knowledgeable source of assistance for using Magellan data through its Magellan Data Products Support Office. This office provides users with information about and assistance in acquiring Magellan data. Standard and special data products are supported, including digital products, photographs, slides, videotapes and NASA Public Information Office products. The office serves NASA-sponsored scientists, other researchers and educators and the general public.

PDS provides an online central catalog showing the general types of Magellan material available and allowing users to identify specific digital and photographic products. Requests for Magellan materials can be made through this central catalog. Additionally, the Geosciences Node provides a detailed catalog of Magellan materials. Users may contact this node directly or through the central PDS catalog. The Geosciences Node supports visiting researchers and provides processing capabilities for using the digital imagery. Contact PDS or the Geosciences Node as follows:

Planetary Data System
Jet Propulsion Laboratory, MS 525-3610
4800 Oak Grove Drive
Pasadena, California 91109-8099
Tel: (818) 306-6130
Fax: (818) 306-6929

E-mail
JPLPDS::PDS_OPERATOR
or
pds_operator@jplpds.jpl.nasa.gov
(137.79.104.100)

Planetary Data System, Geosciences Node
Washington University
St. Louis, Missouri 63130
Tel: (314) 935-5493
Fax: (314) 935-7361

E-mail
WURST::MGNSO
or
mgnso@wurst.wustl.edu
(128.252.135.4)

REGIONAL PLANETARY IMAGE FACILITIES

NASA's Planetary Geology and Geophysics program supports a group of Regional Planetary Image Facilities (RPIFs) around the United States and overseas. The facility at JPL, from where the Magellan mission is directed, is a primary source of Magellan materials. The JPL facility is open by appointment on Monday, Wednesday and Friday to both researchers and members of the public. The JPL facility houses a complete image library of NASA's lunar and planetary missions including Magellan. Visitors can browse through Magellan materials, including the CD-ROM digital imagery collection and videotapes. Contact the JPL facility as follows:

Regional Planetary Image Facility
Jet Propulsion Laboratory, MS 202-301
4800 Oak Grove Drive
Pasadena, California 91109-8099
Tel: (818) 354-3343
Fax: (818) 354-3437

UNITED STATES GEOLOGICAL SURVEY

The United States Geological Survey (USGS) has a program to produce maps of the planets based on the best available data from NASA and foreign missions. NASA-funded researchers can obtain maps directly by contacting the USGS as follows:

Planetary Data Facility
U.S. Geological Survey
Flagstaff, Arizona 86001
Tel: (602) 556-7262
Fax: (602) 556-7090

Any interested persons can obtain a listing of planetary maps from the same address. Persons not funded by NASA can order maps by writing the following:

Branch of Distributions
U.S. Geological Survey
Denver, Colorado 80225
Tel: (303) 236-7477

For more information, also contact the USGS as follows:

Earth Science Information Center
U.S. Geological Survey
Reston, Virginia 22092
Tel: (703) 860-6045

ONLINE COMPUTER ACCESS

In addition to the online computer catalogs provided by NSSDC and PDS, Magellan information and digital images can be obtained through several computer networks. These are described in the following:

SPACELINK, operated by NASA's Marshall Space Flight Center in Alabama, is an electronic information system (a computer bulletin board) oriented toward educators interested in using NASA materials in classes. SPACELINK can be reached either by telephone modem or through the Internet network. Contact —

Telnet: spacelink.msfc.nasa.gov (192.149.89.61)
Modem: (205) 895-0028

NASA's Ames Research Center in California allows public access through the Internet network to information on NASA missions as well as image display software, digital image files in a variety of formats and captions for all press-release images. Magellan images are available in both GIF and VICAR formats. In addition, Magellan CD-ROMs are available in rotation with other NASA image CD-ROMs as a pair of public-access directories. Contact —

ftp: ames.arc.nasa.gov (128.102.18.3)
user: anonymous
cd: pub/SPACE/MAGELLAN, VICAR, GIF, CDROM, CDROM2, SOFTWARE

Finally, the Planetary Data System Geosciences Node at Washington University permits access through the Internet network to selected Magellan data and documentation. Contact —

ftp: wuarchive.wustl.edu (128.252.135.4)
user: anonymous
cd: graphics/magellan

Bibliography

This bibliography provides a listing of references for the The Face of Venus. Magellan-related entries were taken mostly from special issues of the Journal of Geophysical Research (Volume 97, Issues E8 and E10, 1992). These special issues contain additional references. For a detailed bibliography of publications dealing with geology and geophysics of Venus, consult the compilation, Selected Readings on Venus, by S. B. Yewell (1992, Magellan Data Management and Archive Team), which is available upon request from the Regional Planetary Image Facility at JPL:

Regional Planetary Image Facility
Jet Propulsion Laboratory, MS 202-301
Pasadena, California 91109-8099
Tel: (818) 354-3343
Fax: (818) 354-3437

A

Arvidson, R. E., R. Greeley, M. C. Malin, R. S. Saunders, N. R. Izenberg, J. J. Plaut, E. R. Stofan, and M. K. Shepard. "Surface Modification on Venus as Inferred From Magellan Observations of Plains." 1992: J. Geophys. Res. 97, 13,303.

Arvidson, R. E., R. J. Phillips, and N. R. Izenberg. "Global Views of Venus From Magellan." 1992: EOS 73, 161.

Asimow, P. D. and J. A. Wood. "Fluid Outflow From Venus Impact Craters: Analysis From Magellan Data." 1992: J. Geophys. Res. 97, 13,643.

B

Baker, V. R., G. Komatsu, T. J. Parker, V. C. Gulick, J. S. Kargel, and J. S. Lewis. "Channels and Valleys on Venus: Preliminary Analysis of Magellan Data." 1992: J. Geophys. Res. 97, 13,421.

Barker, R. "Venus at Western Elongation." 1934: J. B. A. A. 44, 302.

Barsukov, V. L. and 25 others. "The Geology and Geomorphology of the Venus Surface as Revealed by the Radar Images Obtained by Venera 15 and 16." 1986: Proc. Lunar Sci. Conf. 16th, D378.

Barsukov, V. L., A. T. Basilevsky, V. P. Volkov, and V. N. Zharkov, eds. Venus Geology, Geochemistry, and Geophysics: Research Results From the USSR. 1992: University of Arizona Press, Tucson.

Basilevsky, A. T., R. O. Kuzmin, O. V. Nikolaeva, A. A. Pronin, L. B. Ronca, V. S. Avduevsky, G. R. Uspensky, Z. P. Cheremukhina, V. V. Semenchenko, and V. M. Ladygin. "The Surface of Venus as Revealed by the Venera Landings: Part II." 1985: Geol. Soc. Amer. Bull. 96, 137.

Bindschadler, D. L., A. deCharon, K. K. Beratan, S. E. Smrekar, and J. W. Head. "Magellan Observations of Alpha Regio: Implications for the Formation of Complex Ridged Terrains on Venus." 1992: *J. Geophys. Res.* 97, 13,563.

Bindschadler, D. L., G. Schubert, and W. M. Kaula. "Coldspots and Hotspots: Global Tectonics and Mantle Dynamics of Venus." 1992: *J. Geophys. Res.* 97, 13,495.

Brown, W. E., C. Elachi, R. L. Jordan, A. Laderman, and T. W. Thompson. *Planetary Imaging Radar Study.* JPL Document 701-145 (Internal Document). 1972: Jet Propulsion Laboratory, California Institute of Technology, Pasadena.

Burba, G. A. "Names on the Maps of Venus: A Pre-Magellan Review." 1990: *Earth, Moon, and Planets* 50/51, 541.

Burgess, E. *Venus: An Errant Twin.* 1985: Columbia University Press, New York.

C

Campbell, B. A. and D. B. Campbell. "Analysis of Volcanic Surface Morphology on Venus From Comparison of Arecibo, Magellan, and Terrestrial Airborne Radar Data." 1992: *J. Geophys. Res.* 97, 16,293.

Campbell, D. B., J. W. Head, J. K. Harmon, and A. A. Hine. "Venus: Identification of Banded Terrain in the Mountains of Ishtar Terra." 1983: *Science* 221, 644.

Cattermole, P. *Venus: The Geological Story.* 1994: Johns Hopkins University Press, Baltimore.

Cooper, H. S. F. *The Evening Star: Venus Observed.* 1993: Farrar Strauss Giroux, New York.

D

Davies, M. E., T. R. Colvin, P. G. Rogers, P. W. Chodas, W. L. Sjogren, E. L. Akim, V. A. Stepanyantz, and Z. P. Vlasova. "The Rotation Period, Direction of North Pole, and Geodetic Control Network of Venus." 1992: *J. Geophys. Res.* 97, 13,141.

F

Fimmel, R. O., L. Colin, and E. Burgess. *Pioneer Venus.* NASA SP-461. 1992: National Aeronautics and Space Administration, Washington, D.C.

Ford, J. P., J. J. Plaut, C. M. Weitz, T. G. Farr, D. A. Senske, E. R. Stofan, G. Michaels, and T. J. Parker. *Guide to Magellan Image Interpretation.* JPL Publication 93-24. 1993: Jet Propulsion Laboratory, California Institute of Technology, Pasadena.

Ford, P. G. and G. H. Pettengill. "Venus Topography and Kilometer-scale Slopes." 1992: *J. Geophys. Res.* 97, 13,103.

Friedman, L. D. and J. R. Rose. *Venus Orbital Imagng Radar (VOIR) Study.* JPL Document 760-89 (Internal Document). 1973: Jet Propulsion Laboratory, California Institute of Technology, Pasadena.

G

Gingerich, O. "Galileo and the Phases of Venus." 1984: *Sky Telescope* 68, 520.

Greeley, R. *Planetary Landscapes, 3rd Ed.* 1994: Chapman & Hall, New York.

Greeley, R. and R. M. Batson, Eds. *Planetary Mapping.* 1990: Cambridge University Press, Cambridge.

Greeley, R., R. E. Arvidson, C. Elachi, M. A. Geringer, J. J. Plaut, R. S. Saunders, G. Schubert, E. R. Stofan, E. J. P. Thouvenot, S. D. Wall, and C. M. Weitz. "Aeolian Features on Venus: Preliminary Magellan Results." 1992: *J. Geophys. Res.* 97, 13,319.

Greeley, R., G. Schubert, D. Limonadi, K. C. Bender, W. I. Newman, P. E. Thomas, C. M. Weitz, and S. D. Wall. "Wind Streaks on Venus: Clues to Atmospheric Circulation." 1994: *Science* 263, 358.

Grimm, R. E., and R. J. Phillips. "Anatomy of a Venusian Hot Spot: Geology, Gravity, and Mantle Dynamics of Eistla Regio." 1992: *J. Geophys. Res.* 97, 16,035.

Guest, J. E., M. H. Bulmer, J. Aubele, K. Beratan, R. Greeley, J. W. Head, G. Michaels, C. Weitz, and C. Wiles. "Small Volcanic Edifices and Volcanism in the Plains of Venus." 1992: *J. Geophys. Res.* 97, 15,949.

H

Head, J. W., D. B. Campbell, C. Elachi, J. E. Guest, D. P. McKenzie, R. S. Saunders, G. G. Schaber, and G. Schubert. "Venus Volcanism: Initial Analysis From Magellan Data." 1991: *Science* 252, 276.

Head, J. W., L. S. Crumpler, J. C. Aubele, J. E. Guest, and R. S. Saunders. "Venus Volcanism: Classification of Volcanic Features and Structures, Associations, and Global Distribution From Magellan Data." 1992: *J. Geophys. Res.* 97, 13,153.

Hunt, G. E. and P. Moore. *The Planet Venus.* 1982: Faber & Faber, London.

Hunten, D. M., L. Colin, T. M. Donahue, and V. I. Moroz. *Venus.* 1983: University of Arizona Press, Tucson.

I

Ivanov, B. A. "Venusian Impact Craters on Magellan Images: View From Venera 15/16." 1990: *Earth, Moon, and Planets 50/51,* 159.

Izenberg, N. R. "Venusian Extended Ejecta Deposits as Time-Stratigraphic Markers." 1992: *International Colloq. on Venus, Lunar Planet. Inst. Contrib. 789,* 49.

J

Janes, D. M., S. W. Squyres, D. L. Bindschadler, G. Baer, G. Schubert, V. L. Sharpton, and E. R. Stofan. "Geophysical Models for the Formation and Evolution of Coronae on Venus." 1992: *J. Geophys. Res.* 97, 16,055.

Johnson, W. T. K. "Magellan Imaging Radar Mission to Venus." 1991: *Proc. IEEE 79,* 777.

K

Kaula, W. M., D. L. Bindschadler, R. E. Grimm, V. L.Hansen, K. M. Roberts, and S. E. Smrekar. "Styles of Deformation in Ishtar Terra and Their Implications." 1992: *J. Geophys. Res.* 97, 16,085.

Kuzmin, A. D. and M. Ya. Marov. *Physics of the Planet Venus.* 1974: Nauka, Moskva. [English Translation: NASA-TT-F-16226, National Aeronautics and Space Administration, Washington, D.C.]

L

Lowell, P. "Determination of the Rotation Period and Surface Character of the Planet Venus." 1897: *M. N. R. A. S. 57,*148.

M

Magee Roberts, K., J. E. Guest, J. W. Head and M. G. Lancaster. "Mylitta Fluctus, Venus: Rift-related, Centralized Volcanism and the Emplacement of Large-volume Flow Units." 1992: *J. Geophys. Res.* 97, 15,991.

Masursky, H., E. Eliason, P. G. Ford, G. E. McGill, G. H. Pettengill, G. G. Schaber, and G. Schubert. "Pioneer Venus Radar Results: Geology From Images and Altimetry." 1980: *J. Geophys. Res.* 85, 8232.

McKenzie, D., P. G. Ford, C. Johnson, B. Parsons, D. Sandwell, R. S. Saunders, and S. C. Solomon. "Features on Venus Generated by Plate Boundary Processes." 1992: *J. Geophys. Res.* 97, 13,533.

Moore, H. J., J. J. Plaut, P. M. Schenk, and J. W. Head. "An Unusual Volcano on Venus." 1992: *J. Geophys. Res.* 97, 13,479.

Moore, P. *The Planet Venus.* 1961: Macmillan, New York.

O

Okerson, D., 1992. *Magellan Resources: Access to Magellan Project Information and Science Data.* JPL Document D-9934 (Internal Document). 1992: Jet Propulsion Laboratory, California Institute of Technology, Pasadena.

P

Pavri, B., J. W. Head, K. B., Klose, and L. Wilson. "Steep-sided Domes on Venus: Characteristics, Geologic Setting, and Eruption Conditions From Magellan Data." 1992: *J. Geophys. Res.* 97, 13,445.

Pettengill, G. H., P. G. Ford, and B. C. Chapman. "Venus: Surface Electromagnetic Properties." 1988: *J. Geophys. Res.* 93, 14,881.

Pettengill, G. H., P. G. Ford, W. T. K. Johnson, R. K. Raney, and L. A. Soderblom. "Magellan: Radar Performance and Data Products." 1991: *Science* 252, 260.

Pettengill, G. H., P. G. Ford, and R. J. Wilt. "Venus Surface Radiothermal Emission as Observed by Magellan." 1992: *J. Geophys. Res.* 97, 13,091.

Phillips, R. J., R. E. Arvidson, J. M. Boyce, D. B. Campbell, J. E. Guest, G. G. Schaber, and L. A. Soderblom. "Impact Craters on Venus: Initial Analysis From Magellan." 1991: *Science* 252, 288.

Phillips, R. J. and V. L. Hansen. "Tectonic and Magmatic Evolution of Venus." 1992: *Annu. Rev. Earth Planet. Sci.* 22, 597.

Phillips, R. J., R. F. Raubertas, R. E. Arvidson, I. C. Sarkar, R. R. Herrick, N. Izenberg, and R. E. Grimm. "Impact Craters and Venus Resurfacing History. 1992: *J. Geophys. Res.* 97, 15,923.

S

Saunders, R. S., G. H. Pettengill, R. E. Arvidson, W. L. Sjogren, W. T. K. Johnson, and L. Pieri. "The Magellan Venus Radar Mapping Mission." 1990: *J. Geophys. Res.* 95, 8339.

Saunders, R. S., R. E. Arvidson, J. W. Head, G. G. Schaber, E. R. Stofan, and S. C. Solomon. "An Overview of Venus Geology." 1991: *Science* 252, 249.

Saunders, R. S. and G. H. Pettengill. "Magellan: Mission Summary." 1991: *Science* 252, 247.

Schaber, G. G., R. G. Strom, H. J. Moore, L. A. Soderblom, R. J. Kirk, D. J. Chadwick, D. D. Dawson, L. R. Gaddis, J. M. Boyce, and J. Russell. "Geology and Distribution of Impact Craters on Venus: What Are They Telling Us?" 1992: *J. Geophys. Res.* 97, 13,257.

Schultz, P. H. "Atmospheric Effects on Ejecta Emplacement and Crater Formation on Venus From Magellan." 1992: *J. Geophys. Res.* 97, 16,183.

Senske, D. A., G. G. Schaber, and E. R. Stofan. "Regional Topographic Rises on Venus: Geology of Western Eistla Regio and Comparison to Beta Regio and Atla Regio." 1992: *J. Geophys. Res.* 97, 13,395.

Shepard, M. K., R. E. Arvidson, R. A. Brackett, and B. Fegley, Jr. "A Ferroelectric Model for the Low Emissivity Highlands on Venus." 1993: *Geophys. Res. Lett.* (submitted).

Solomon, S. C. and J. W. Head. "Fundamental Issues in the Geology and Geophysics of Venus." 1991: *Science* 252, 252.

Solomon, S. C., S. E. Smrekar, D. L. Bindschadler, R. E. Grimm, W. M. Kaula, G. E. McGill, R. J. Phillips, R. S. Saunders, G. Schubert, S. W. Squyres, and E. R. Stofan. "Venus Tectonics: An Overview of Magellan Observations." 1992: *J. Geophys. Res.* 97, 13,199.

Solomon, S. C. "The Geophysics of Venus." 1993: *Physics Today* 46, 48.

Squyres, S. W., D. M. Janes, G. Baer, D. L. Bindschadler, G. Schubert, V. L. Sharpton, and E. R. Stofan. "The Morphology and Evolution of Coronae on Venus." 1992: *J. Geophys. Res.* 97, 13,611.

Squyres, S. W., D. G. Jankowski, M. Simons, S. C. Solomon, B. H. Hager, and G. E. McGill. "Plains Tectonism on Venus: The Deformation Belts of Lavinia Planitia." 1992: *J. Geophys. Res.* 97, 13,579.

Stofan, E. R., V. L. Sharpton, G. Schubert, G. Baer, D. L. Bindschadler, D. M. Janes, and S. W. Squyres. "Global Distribution and Characteristics of Coronae and Related Features on Venus: Implication for Origin and Mantle Processes." *J. Geophys. Res.* 97, 13,347.

T

Tyler, G. L., P. G. Ford, D. B. Campbell, C. Elachi, G. H. Pettengill, and R. A. Simpson. "Magellan: Electrical and Physical Properties of Venus Surface." 1991: *Science* 252, 265.

Tyler, G. L., R. A. Simpson, M. J. Maurer, and E. Holman. "Scattering Properties of the Venusian Surface: Preliminary Results From Magellan." 1991: *J. Geophys. Res.* 97, 13,115.

W

Weitz, C. M. and A. T. Basilevsky. "Magellan Observations of the Venera and Vega Landing Site Regions." 1993: *J. Geophys. Res.* 98, 17,069.

Y

Yewell, S. B. *Magellan Data Product Information Handbook.* JPL Document D-11020 (Internal Document). 1993: Jet Propulsion Laboratory, California Institute of Technology, Pasadena.

Young, C., Ed. *The Magellan Venus Explorer's Guide.* JPL Publication 90-24. 1990: Jet Propulsion Laboratory, California Institute of Technology, Pasadena.

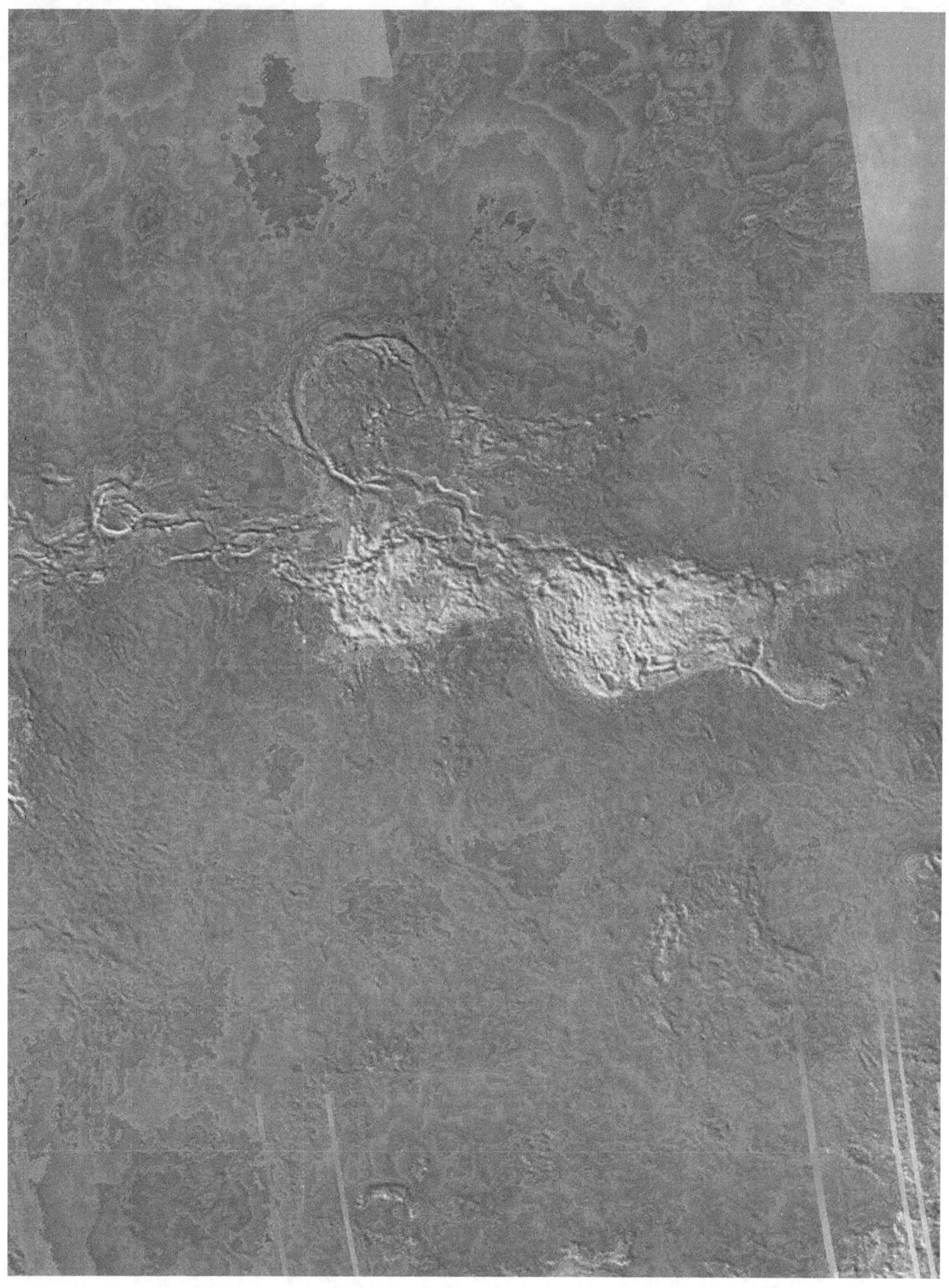

www.ingramcontent.com/pod-product-compliance
Lightning Source LLC
Chambersburg PA
CBHW081454170526
45166CB00008B/2423